Fisheries Science

R. Santhanam

2013
DAYA PUBLISHING HOUSE
A Division of
Astral International Pvt. Ltd.
New Delhi - 110 002

First Published: 1990
Reprint: 2013

ISBN 978-81-7035-085-9

Published by : **Daya Publishing House**
A Division of
Astral International Pvt. Ltd.
81, Darya Ganj, Near Hindi Park,
Delhi Medical Association Road,
New Delhi - 110 002
Phone: 011-4354 9197, 2324 8134
Fax: +91-11-2324 3060
E-mail: info@astralint.com
Website: www.astralint.com

Laser Typesetting : **Classic Computer Services**
Delhi - 110 035

Printed at : **Salasar Imaging Systems**
Delhi - 110 035

PRINTED IN INDIA

Foreword

Culture and capture fisheries have assumed greater significance in India in recent years in the context of the growing demand for fish, for mitigating the problems of malnutrition, unemployment and gaps in foreign trade. It is necessary to intensify research and development in fisheries to help achieve the targeted fish production of about five million tonnes by 2000 AD. The focus should be on the optimum exploitation of the hitherto underexploited fisheries resources in the exclusive economic zone of our country and on aquaculture.

One of the major responsibilities of fisheries professionals and researchers is to consolidate the information base into standard publications for the benefit of the fishing industry trade, aquaculturists, policy makers and development departments. The growing information base in fisheries including Aquaculture, Fisheries Environment, Fisheries Biology and Fisheries Technology is very wide, scattered in dozens of publications of various types. Keeping this in view, the author has prepared this compendium which is a compilation mainly of his published articles.

I am confident that this publication shall serve as a useful manual to all those interested in fisheries.

Dr. P. RICHARD MASILLAMONY
Vice Chancellor
Tamil Nadu Veterinary and
Animal Sciences University
Madras

Foreword

[illegible]

[illegible]

[illegible]

[illegible]

Preface

The annual production of fish in India is about 1.5 million tons including 0.14 million tons through aquaculture and it is estimated that about 10 million tons of fish and shellfish may be required to meet the demand of the increasing human population. It is, therefore, felt that new techniques have to be developed for the production of fish through aquaculture and capture fisheries besides creating an awareness among public about Fisheries Science.

Although several books on the individual aspects of Fisheries Science are available from abroad, a comprehensive compendium incorporating modern techniques relating to various aspects of Fisheries Science with special reference to Indian conditions is lacking. In order to fill up this long felt gap, this publication is being broughtout.

The present publication, a compilation of mainly published articles in leading English Dailies and Magazines has three parts. The first part deals with important culture practices relating to freshwater and brackishwater systems. The second part deals with articles relating to the distribution and abundance of fish and invertebrates and the third part deals with modern techniques of marine capture fisheries and fish and shellfish processing and production of value added fisheries products. All these articles have been written in an easy to read style with suitable illustrations and I hope that the publication would serve as a valuable guide for fisheries students, aquaculturists, fisheries technologists, marine biologists and general public interested in fisheries.

I am grateful to my colleagues Dr. G. Jegatheesan, Dr. K. Venkataramanujam and Dr. V. Sundararaj (Professors); Dr. P. Natarajan formerly Associate Professor of this College; Mr. G. Sanjeeviraj, Dr. N. Sukumaran and Dr. N. Ramanathan (Associate Professors); Mr. P. Gopalakrishnan, Smt. Indra Jasmine, Mr. M. Venkatasamy, Mr. R. Raghavan, Mr. P. Velayutham, Mr. R. Senthilathiban, Mr. A. Srinivasan and Mr. B. Manimaran who were the coauthors of the published articles.

I am grately indebted to Dr. P. Richard Masillamony, Vice Chancellor, Tamil Nadu Veterinary and Animal Sciences University for according permission to publish this compendium and for Foreword. I sincerely thank Dr. M. Devaraj, Dean, Fisheries College for suggestions and encouragements. Grateful acknowledgements are also made to The Hindu, Indian Express, Science Reporter, Fishing Chimes, Shipping and Marine Industries Journal, Seafood Export Journal, Kisan World, Aquaworld and Industrial Fisheries Association's

Annual wherein some of the articles incorporated in this compilation were published earlier and to the Murugappa Chettiar Research Centre, Tharamani, Madras for information on Fish Aggregating Devices. I also thank Miss. R. Ezhilrani, Library Assistant for bibliographical assistance and Mr. S. Madhavan, Assistant for typing the manuscript.

Suggestions from the readers are welcome.

The Author

Contents

Foreword III
Preface V

I. Aquaculture
1. Aquaculture - Hope for Combating Malnutrition 1
2. Composite Fish Culture 4
3. Integrated Fish Farming is Lucrative 7
4. Can Sewage be Profitably Utilised? 10
5. Fish Farming Using Sewage Wastes 15
6. Freshwater Prawn-cum-carp Farming - A New Polyculture Practice 17
7. Tilapia-Tarpon Culture in Fresh and Brackishwaters 21
8. Milkfish - Prawn Farming in Brackishwaters 24
9. Crab Culture in Coastal Ponds 28
10. Oyster Farming in Brackishwaters 31
11. Pole Farming of Edible Oysters in Brackishwaters 33
12. Scallop Culture 36
13. Marine Mussels 39
14. Culture of Seaweeds in Brackishwaters - A New and Promising Technology 43
15. Bloodworms - Their Culture and Prospects 46
16. Water Recirculation Unit for Profitable Fish Culture 49
17. Hydroponics and Fish Culture 51
18. Non-Conventional Feeds for Profitable Aquaculture 53
19. HCG Induced Breeding in Freshwater Fishes - A Boon to Fish Farmers 55
20. Guppy - A Unique Ornamental Fish 58
21. Biological Filters and Air-lift Pumps for Aquaria 62
22. Balanced Diets for Aquarium Fishes 65
23. How to Protect Aquarium Fishes from Diseases 69
24. Mass Culture of Zooplankton for Coastal Aquaculture 74
25. New Techniques in the Transport of Fish and Prawn Seed for Aquaculture 76
26. New Methods of Preservation of Fish Gametes for Aquaculture. 80

II. Marine Biology

27. Artificial Seawater 82
28. Sea as a Source of Drugs 84
29. Economic Importance of Diatoms 86
30. Plankton and its Relation to Fisheries 88
31. Tintinnids (Marine Protozoans) as Fishery Indicators 94
32. Rotifers as Indicators of Water Quality and Pollution 97
33. Planktonic Molluscs as Indicators of Ocean Currents 100
34. Menace of the Marine Foulers and Borers 103
35. Beautiful Corals 106
36. Oceanic Insects 112
37. The Pistol Shrimps 114
38. Sea Cucumbers are Rich in Protein 116
39. Why Should We Eat Fish? 118
40. Why Fish Smells? 120
41. Puffer-The Most Dangerous Fish 121
42. Fishes that Shed Skin 123
43. Unique Devices to Study Fish Behaviour in Polluted Area 125

III. Fisheries Technology

44. Electricity in Fishing 128
45. An Electronic Device for Detecting Fishes 134
46. Devices to Lure Fish for Bumper Catch 139
47. Solar Driers for Hygienic Drying of Fish and Farm Produce 141
48. Prawn Picking - A Sustenance for Rural Women Folk 143
49. Fishery Byproducts of Commerce 145
50. Industrial Uses of Prawn Shell Wastes 150
51. Fish Sauce - A New and Promising Byproduct of Commerce 152
52. Ambergris 154
53. Multifarious Uses of Algae 155
54. Prospects of Indian Seaweeds 158
55. Seaweeds as Fertilisers 161
56. Dunaliella - A Unique Halophilic Microalga 162
57. Profitable Uses of Freshwater Weeds 164

Index 167

Legends to Figures

1. Fish catch from a freshwater fish pond
2. The different types of fish in the same pond to maximise yield
3. Integrated farming - foodweb
4. Plan of sewage fish culture unit suitable for an educational Institution
5. Plan of sewage fish culture unit suitable for local bodies
6. Schematic diagram of sewage fish culture unit
7. *Macrobrachium rosenbergii*
8. *M. malcomsonii*
9. Tilapia
10. Tarpon
11. Milkfish
12. *Penaeus indicus*
13. *P. monodon*
14. *Scylla serrata*
15. *Crassostrea madrasensis*
16. Oysters grown in pole
17. Oysters grown in bamboo pole
18. *Pecten plica*
19. *P. tranquebaricus*
20. *P. crassicostatus* (outer view)
21. *P. crassicostatus* (inner view)
22. *P. splendidulus*
23. *Perna viridis*
24. *P. indica*
25. Raft for mussel culture
26. *Ulva*

    a. Bamboo pole; b. Float; c. Culture rope; d. Growing mussel;
    e. Sinker; f. Anchor
27. *Enteromorpha*
28. *Caulerpa*
29. *Sargassum*
30. *Turbinaria*
31. *Gracilaria*
32. *Gelidiella*
33. Bloodworm

34. Water recirculation unit
 a. Lower reservoir; b. Bump; c. Upper reservoir; d. Level maintainer; e. UV light; f. Biological filter; g. Cascader; h. Fish tank
35. Hydroponic system for plant-fish culture
 a. Airlifted water; b. Compressed air; c. Growing plant; d. Growing fish; e. Gravel (biological filter); f. False bottom
36. Administering intramuscular injéction of HCG pituitary hormone to a carp
37. Breeding hapa
38. Hatching hapa
39. (a-r) Different varieties of guppy based on tail variations
40. Male guppy
41. Female guppy showing brood incubating pouch
42. Female guppy showing brood with fully matured eggs
43. Female guppy in the breeding trap
44. Biological filter. 1. Compressed air; 2.Gravel; 3. False bottom
45. Air lift
46. Air lift pump with in-built biological filter unit.
 a. Compressed air; b. Water level; c. Gravel; d., Air diffuser
47. Hand pelletiser
48. A platy affected by tail or fin rot
49. Life cycle of *Ichthyophthirius*
50. *Costia necatrix*
51. A danio affected by costiasis
52. *Gyrodactylus*
53. *Dactylogyrus*
54. *Lernaea*
55. *Lernaenicus*
56. A portion of fish affected by *Lernaenicus*
57. *Caligus*
58. Gills of a fish affected by calligid parasites
59. *Argulus*
60. An angel fish affected by leeches
61. *Moina*
62. *Daphnia*
63. Oxygen cylinder and tin for fish seed transport
64. Method of oxygen packing for fish seed transport
65. Marine diatoms. a. *Coscinodiscus; b. Planktoniella; c. Stephanopyxis; d. Rhizosolenia; e. Climacodium; f. Biddulphia; g. Hemidiscus; h. Thalassiothrix; i. Gyrosigma*
66. Features of a tintinnid
67. *Rhabdonella*
68. *Codonellopsis*
69. *Tintinnopsis nordgvisti*
70. *Coxliella annulata*
71. *Favella philippinensis*

72. *Amphorellopsis acuta*
73. *T. mortensenii*
74. *Metacylis jorgensenii*
75. *T. sacculus*
76. *T. tocantinensis*
77. *Favella philippinensis*
78. *Coxliella ampla*
79. *T. cylindrica*
80. *T. tubulosa*
81. *Dictyocysta seshaiyaii*
82. *Eutintinnus tenuis*
83. *Helicostomella longa*
84. *Brachionus angularis*
85. *B. calciflorus*
86. *B. rubens*
87. *Filina longiseta*
88. *Rotatoria neptunia*
89. *R. rotatoria*
90. *B. urceolaris*
91. *Keratella cochlearis*
92. *Lepadella ovalis*
93. *Polyarthra vulgaris*
94. *Lecane bulla*
95. *B. plicatilis*
96. *Hexarthra fennica*
97. *B. quadridentatus*
98. *Lecane lunaris*
99. *Trichocerca similis*
100. *Janthina*
101. *Limacina*
102. *Creseis*
103. *Cavolinia*
104. *Clione*
105. Barnacle nauplius
106. Barnacle cypris
107. Barnacle adult
108. *Limnoria*
109. *Sphaeroma*
110. Shipworm
111. Section of fringing reef
112. Section of barrier reef
113. Section of an atoll
114. Coral polyps
115. Section of a portion of coral colony and its skeleton
116-117. Some species of corals of Tuticorin coast
118. *Halobates micans*

119. Eggs of *Halobates*
120. Newly hatched nymph of *Halobates*
121. An adult pistol shrimp
122. A large chela or pistol hand
123. Sea cucumber
124. Pufferfish
125. *Congiopodus spinifer*
126. *Acanthosphex leurynnis*
127. Moulted skin of *A. leurynnis*
128. Blazka apparatus in operation
129. An entire view of Blazka apparatus
130. Fry respirometer
131. Fish behaviour in actual fields of fishing gear
132. Pump fishing with light and electricity
133. Pump fishing with light for squid of the Californian coast
134. Electric seines 1. Bag; 2. Seine wing; 3. Linear electrode
135. Employment of electrodes of lift nets in sardine fishing.
1. Electrodes; 2. Lamps; 3. Focusing lamps.
136. Bottom trawl for inland water bodies (East Germany)
1. Draw bar; 2. Cathode; 3. Anode wire fixed to upper trawl panel
137. Electrical pair trawl for inland water bodies
1. Electrode (flexible metallic cable whipping); 2. Fleat; 3. Underwater pulse generator
138. Shrimp trawl with self contained power supply
1. Pulse generator; 2. Feed cable; 3. Adjusting rope; 4. Electrodes.
139. Saury fishing with the light-electric field of current-pump method.
1. Water separator; 2. R.B. 150 fish pump; 3. Suction hose;
4. Check valve; 5. Anode; 6. Buoy; 7.Focusing light source;
8. Counterweight; 9. Hoisting cables; 10. Cathodes;
11. Rod shaped anode.
140. Fishing for black sea anchovy. 1. Spherical anode; 2. Source of light;
3. Suction hose; 4. Check valve; 5. Fish pump; 6. Water separator.
141. Different parts of an Echosounder
142. Influence of pen speed in echogram
143. Influence of paper speed in echogram
144. Effect of pulse length in echogram
145. Effect of phasing
146. Importance of suppression in echogram
147. Echograms with and without whiteline
148. Fish aggregating device
149. Solar tent drier
150. *Dunaliella salina*
151. A flow chart showing utilisation of freshwater weeds.

1

Aquaculture—Hope for Combating Malnutrition

India's Annual Fish Production is 2.1 million tonnes and the share of aquaculture is only 0.14 million tonnes. The annual per capita fish consumption of our country is also very low (3 kg) compared to the world average of 12 kg. The advisory committee on human nutrition has on the other hand recommended 31 kg. This shows that the demand for fish outstrips supply and aquaculture is considered to be the only alternative to capture fisheries for producing more protein to combat malnutrition.

Aquaculture, the culture of the commercially important animals, is a dependable year round source of animal protein unlike agriculture or animal husbandry. Further, it is a more profitable venture than maintaining live-stock as the food conversion ratio in fish is better than pig, chick or cattle. Aquaculture currently provides 10% of world's water derived protein and about 3% of all world's protein exclusive of milk.

Freshwater aquaculture: Though freshwater bodies of India amount to 7.5 million ha, we at present use only 1.7 million ha of reservoirs and lakes and 0.15 million ha of ponds for an annual aquaculture yield of 0.14 Million tonnes. Hence more freshwater bodies have to be brought under aquaculture using the following practices.

Composite fish culture: Fast-growing compatible carps with different feeding habits such as catla, rohu, mrigal, silver carp, grass carp and common carp can be stocked together in earthen ponds and get an yield of more than 10 t/ha. (Fig. 1)

Air-breathing fish culture: Derelict shallow waters (swamps) with poor oxygen content could be utilised for culturing air-breathing fishes such as maghur, singhi and murrels. In India, about 0.4 million hectare of swampy areas are available for this practice.

Sewage-fed fish culture: Sewage waters can be treated in oxidation ponds and subsequently used for fish culture. This technology can be availed of in big cities like Bombay, Madras, Bangalore, etc. where part of sewage is now drained into adjoining wet lands. These wet lands can be used for sewage-fed fish culture. As there is no expenditure towards fertiliser and feed, this is a very lucrative venture and an yield of 8-13 t/ha/yr is possible besides making the environment clean.

Integrated fish farming: Fish can be cultured alongwith agricultural crops such as rice and banana and live-stock such as poultry, duck, cattle and pig.

Courtesy : Shipping and Marine Industries Journal

While rice and banana fields give space for fish culture, live-stock excreta is either recycled or serve as direct food for fish. The pig-cum fish culture and duck-cum fish culture have given an yield of 7000 kg/ha and 4000 kg/ha respectively. Pig-cum fish culture can also benefit tribal population in North-eastern hill states where pigs are traditionally reared. Similarly, duck-cum fish culture can be practised in Tamil Nadu, Kerala, Assam, Orissa, West Bengal and Bihar where duck rearing is common.

Fig. 1

Fish culture in rice fields: Fish like carps, murrels and tilapia can be cultivated simultaneously or in rotation in the trenches adjoining rice fields which offer additional revenue to the agriculturists. It is possible to get a fish yield of 700 kg/ha during the paddy cultivable seasons. Though an area of 6 million ha is under rice cultivation in India, only 0.03% of this is now used at present for rice-fish culture. Hence, there is a good scope for implementing this venture as an agro-based industry especially in the rainfed low-lying areas covering an estimated area of 2 million ha in West Bengal, Assam, Tripura, Orissa, Bihar and U.P.

Larvivorous fish culture: Larvicidal fishes such as *Oryzias melastigma, Aplocheilus blochi* and *Gambusia affinis* can be cultivated in cement cisterns and their mass production can control malaria, filaria, dengue and brain fever which affect humans through vectors like mosquitoes.

Ornamental fish culture: As aquarium fishes have a good export market in Singapore, Malaysia, UK and USA, the culture and proliferation of these fishes through indoor breeding is a profitable venture.

Brackishwater aquaculture: Our country is also blessed with 1.7 million ha of brackishwater bodies such as estuaries, backwaters and mangrove waterways. The brackishwater areas such as pokkali fields of Kerala (5117 ha), Bheries of West Bengal (20000 ha), Gazani farms of Karnataka and Khozan lands of Goa have been found more suitable for undertaking the traditional fish and prawn culture practice viz. trapping and filtration. However, large scale farming of prawns is practised only in the brackishwaters of West Bengal, Kerala, Karnataka and Gujarat, covering an area of about 30,000 ha and the present annual yield is about 10,000 tonnes. As brackishwater areas of our country are abundant with over 26 species of commercially important fishes, prawns, crabs and molluscs, they offer vast scope for cultivating fish and prawns. Prawn farming among others is certainly a lucrative venture as it gives an annual production of 2.1 t/ha through three crops. It is also worth mentioning that if the remaining unutilized brackishwater areas are brought under prawn farming, it is possible to produce annually 0.1 million tonnes of prawn which can give a return of about Rs. 500 crores. As prawns are our major item of exports among seafoods (90 per cent earnings) prawn farming especially in brackishwater ponds would not only enhance the national income but create employment opportunities.

2

Composite Fish Culture

The fish pond is a complex ecosystem having various types of food in the different regions. For example, the surface is occupied by the floating live matter such as phyto and zooplankton; the column region is having live and dead matter sunk from the surface; the bottom is enriched with detritus or dead organic matter and the margin of the ponds is invariably harboured by the abundance of unwanted aquatic weeds which is always a nuisance for fish culture operations.

Unless these different trophic levels of fish ponds are utilised, fish culture may not be a profitable venture. The "composite fish culture" a newly formulated practice after intensive research could solve this problem and help to increase fish production.

The composite fish culture or mixed farming is the culture of compatible species of fish. The objective of this type of fish culture is to select and grow fishes of different feeding habits in order to exploit all types of food available in the different regions of pond.

Fishes employed in Composite fish culture: In this culture operation, fast growing fishes namely Indian major carps Catla (*Catla catla*), Rohu (*Labeo rohita*), Mrigal (*Cirrhina mrigala*) and the exotic carps such as Silver carp (*Hypophthalmichthys molitrix*), common carp (*Cyprinus carpio*) and grass carp (*Ctenopharyngodon idella*) are usually employed.

The chief feature of the different species of culturable fishes are:

Catla: It is a surface feeder due to its upturned mouth; feeds chiefly on animal plankton (zooplankton); reaches a maximum size of 1.8 m.

Rohu: It has a terminal mouth with fringed lip to aid column feeding on planktonic algae sunk from the surface; grows over 1m.

Mrigal: It feeds on detritus in the bottom: mouth is slightly shifted ventrally: reaches a maximum size of 90 cm.

Silver Carp: It has an upturned mouth; feeds mainly on the phytoplankton of the surface; attains a size of 60 cm.

Common Carp: It is a bottom feeder on all types of food available in the bottom; eggs adhesive to submerged vegetation

Grass carp: It is a good biological tool for aquatic weed control; feeds on aquatic weeds such as *Hydrilla, Lemna, Spirodella* and *Najas* at a rate of 8 times of its body weight per day. (Fig. 2)

Courtesy : The Hindu

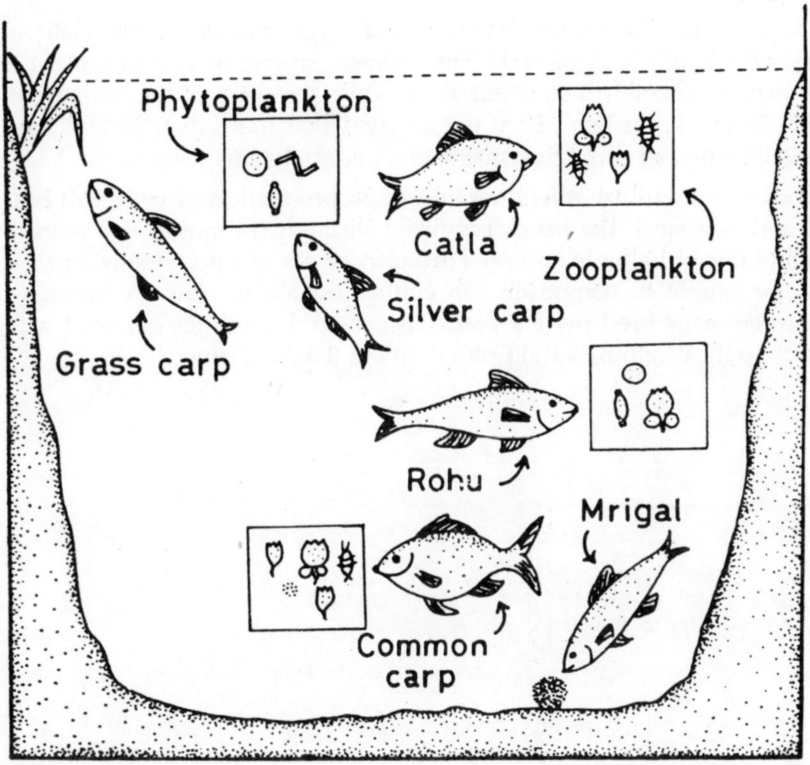

2

Operational Procedures

The optimum size of the fish pond for composite fish culture varies from 0.1 to 0.5 ha. The pond, cleared from aquatic weeds and weed fishes, is to be fertilised. The initial dosage of cowdung is to be 1/6 of the total quantity (10-20 tons/ha/yr) and the rest are applied in equal fortnightly instalments. This is followed by the application (per hectare per year) of inorganic fertilisers which comprise urea at 200 kg/ha/yr or Ammonium sulphate at 450 kg; superphosphate at 250 kg and muriate potash at 40 kg.

The inorganic fertilisers are applied in equal fortnightly instalments alternating with organic fertilisers. Fifteen days after the initial fertilisation fingerlings of catla, rohu, mrigal, silvercarp, commoncarp and grasscarp (with the ratio of 15%, 10%, 20%, 30%, 20% and 5% respectively) are stocked at the rate of 5,000 - 10,000 Nos./ha. Prior to stocking, the fingerlings are to be given a bath of 2% Potassium permanganate to avoid infection.

The stocked fingerlings which usually feed on the natural food namely phyto and zooplankton produced by fertilisation are also fed with supplementary diets of groundnut oilcake and rice bran in the ratio of 1:1 and at the rate of 2-5% body weight of fish stocked in the pond.

For grasscarp, forage grass or even kitchen vegetable refuge may be given. The ponds thus stocked with fingerlings which would have grown to table size are to be harvested after a year.

By adopting the composite fish culture technique, an average fish yield of 8 to 10 tons/ha/yr could be achieved. The average expenditure to produce 1 kg of fish has been estimated to be about Rs. 4 while the price in the market will be about Rs. 15 per kg of fish. So it is estimated that more than 50% of the total income will be the net profit to farmers who undertake this venture.

Composite fish culture offers not only high production of table fish but also good brood fish since the fishes produced through composite fish culture are healthy and fast growing in nature. Furthermore, an additional source of income can also be sought in composite fish culture especially through commoncarp, which can be easily bred twice a year. The sale of fingerlings obtained will add very much to the economics and profitability to the fish farmers.

3

Integrated Fish Farming is Lucrative

Fish culture in combination with livestock and crop-integrated farming is a unique and lucrative venture and provides a higher farm income, availability of cheap source of protein for the rural dwellers, higher productivity among small landholders and an increase in the supply of feeds for the farmers' livestock. (Fig. 3)

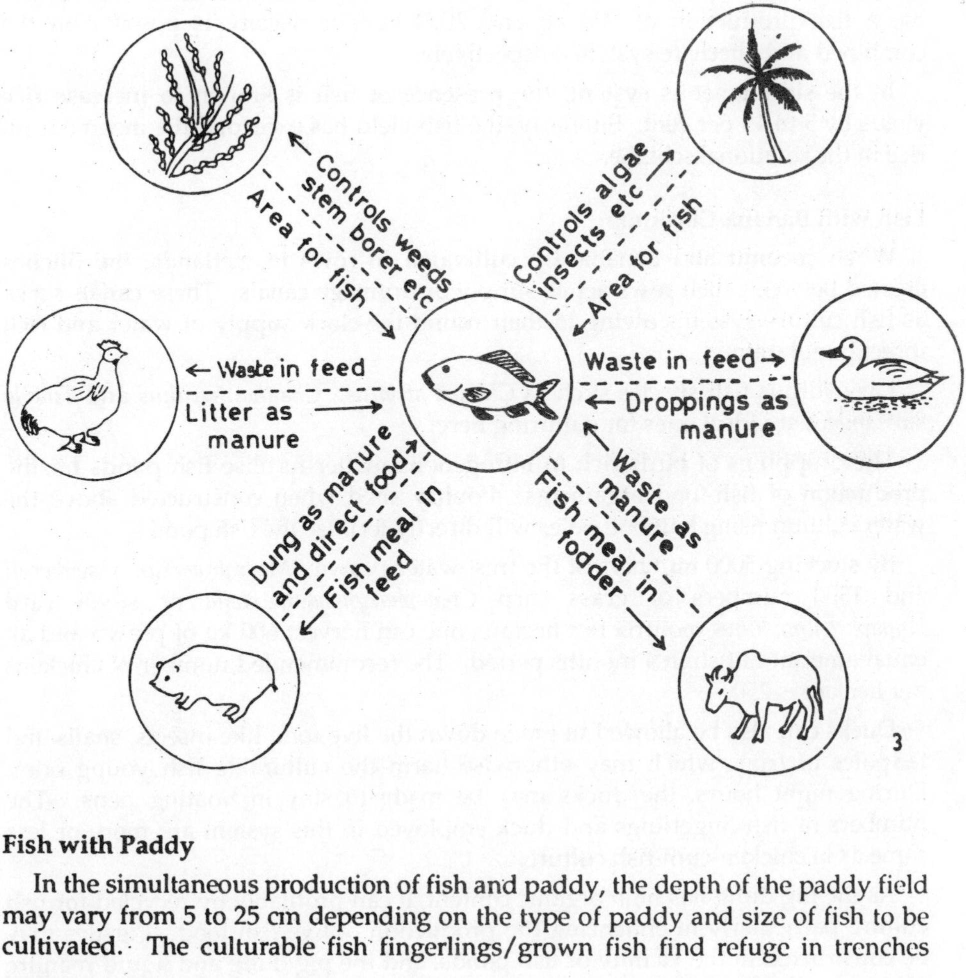

3

Fish with Paddy

In the simultaneous production of fish and paddy, the depth of the paddy field may vary from 5 to 25 cm depending on the type of paddy and size of fish to be cultivated. The culturable fish fingerlings/grown fish find refuge in trenches

---

Courtesy : The Hindu

(ditches) dug in the paddy fields when they are not foraging among rice plants. These consist of a single ditch across the centre of the plot comprising 10 per cent of the paddy production area and a peripheral ditch of 25 per cent of the area.

These ditches having approximately a depth of one metre also serve as capture channels in which the grown fish collect when water is lowered due to seasonal changes. Under the alternate production of paddy, the field after the harvest is converted into a temporary fish pond.

The fish species which can be cultured in the paddy fields must be able to withstand shallow water, high temperature, low oxygen and turbid water. Species such as *Channa striaitus, Channa marulius, Tilapia mossambica, Anabas testudineus, Catla catla, Labeo rohita Cyprinus carpio* and *Clarius macrocephalus* which eat the undesirable insects (including mosquito larvae) of paddy fields are more suitable for this type of farming. Under a stocking density of 2000 fingerlings/ ha, a fish production of 500 kg and 2000 kg/per hectare is possible in the combined and alternate systems respectively.

In the simultaneous system, the presence of fish is known to increase rice yields by 5 to 15 per cent. Similarly, the fish yield has exceeded the income from rice in the rotational scheme.

Fish with Banana/Coconut

When coconut and banana are cultivated in rows in wetlands, the ditches erected between such rows act as supply or drainage canals. These canals serve as fish culture systems owing to their round-the-clock supply of water and rich insect population.

Larvivorous fish species such as *Channa striatus, Channa marulius* and *Tilapia mossambica* are ideal ones for culturing here.

The droppings of birds rich in nitrogenous matter fertilise fish ponds for the production of fish-food organisms. Poultry shed when constructed above the water column using bamboo poles will directly fertilise the fish ponds.

By stocking 5000 numbers of the freshwater prawn. *Macrobrachium rosenbergii* and 1500 numbers of grass carp *Ctenopharyngodon idella* or silver carp *Hypophthalmichthys* molitrix per hectare, one can harvest 600 kg of prawn and an equal amount of fish in 4 months period. The recommended number of chickens per hectare is 250.

Ducks can also be allowed to graze down the live food like insects, snails and tadpoles of frogs which may otherwise harm the culturable fish young ones. During night hours, the ducks may be made to stay in floating pens. The numbers of fish fingerlings and duck employed in this system are more or less same as in chicken-cum-fish culture.

As the pig dung has high organic content, it can profitably by recycled for fish culture particularly in enhancing the production of live fish-food. Pigsties may be constructed in the vicinity of fish ponds, and the pig dung and liquid manure may be washed into the fish ponds.

The optimum dose of the pig manure per hectare has been estimated as 5 tonnes. Based on this, 50 pigs per hectare are recommended. Grass carp, silver carp, tilapia and pig-dung feeding common carp are ideal fish varieties for integration and a production of 3.6 tonnes/ha per year has been obtained under a stocking density of 5000 fingerlings/ha.

Fish with Cattle

In this system, cowbyres are built in the nearby area of fish ponds, so that, cowdung could be transported to the fish ponds at specified intervals. It is thoroughly a productive field for implementation.

4

Can Sewage be Profitably Utilised ?

Most of us think that sewage water spreads obnoxious smell and serves as a breeding ground for mosquitoes and other disease causing organisms. (In many countries sewage water is processed and used for drinking purposes and also for industrial and agricultural purposes such as in culturing fish and cultivating field crops. Fish are known to convert efficiently biowaste into edible protein. Germany, USSR, Poland, Hungary, Israel, China and Indonesia have already obtained an annual fish production of 3.5 to 4.0 tonnes/ha/annum by using domestic sewage as a source of fertiliser in freshwater fish ponds. Fish culture in sewage systems involves meagre investment but the return is substantial. In India the quantity of wastewater discharged daily after human and industrial use has been estimated at 3000 million litres, but less than 2% is used for fish culture. There are only 132 sewage-fed fish farms covering an area of 12,000 ha in India. Sewage-fed fish culture as a commercial venture is undertaken in West Bengal only.

Quality of Sewage Waters

The significance of using sewage in fish culture is due to the fact that it is rich in nutrients and minerals. Raw sewage contains complex organic and inorganic materials such as proteins, urea, cellulose, fats, carbohydrates and soaps. In biological sewage treatment, bacteria, fungi, zooplankton and algae degrade and use these complex materials resulting in an effluent rich in potassium, nitrogen, phosphorus and other elements which are highly useful for fish culture. Recent researches have proved that water hyacinth (*Eichhornia crassipes*) an aquatic weed can efficiently recover the above mentioned elements. During transformation, gases like hydrogen sulphide, methane and ammonia are released and their presence is unsuitable for fish culture. Hence it is necessary to stabilise sewage waters.

Stabilisation of Sewage Waters

If sewage water is stagnated in bigger ponds, otherwise called oxidation ponds, it is loaded with oxygen due to the development of photosynthetic microorganisms particularly phytoplankton. Nutrients required for phytoplankton growth are made available by the bacteria which disintegrate organic substances of sewage waters into useful inorganic substances. Sewage water can be transformed into fertile water by stagnating them for about 10 days. The fish ponds should invariably be half the size of oxidation ponds.

Courtesy : The Hindu

Table 1: **Stocking Ratio for Sewage-fed Fish Ponds**

S. No.	*Variety of fish*	*% of 6 variety Stocking*	*% of 5 variety Stocking*
1.	Silver carp	25	—
2.	Catla	15	40
3.	Rohu	10	10
4.	Mrigal	25	20
5.	Common carp	20	20
6.	Grass carp	5	—
7.	Scale carp	—	10

Table 2: **Cost of Construction of 2 Earthen Ponds of 1000 m³ each (at the rate of Rs. 7.50/m³): Rs. 15,000**

	Annual Expenditure Rs.
Interest at the rate of 10% on initial investment on 2 earthen ponds	1,500.00
Fish seed cost for 2000 fingerlings	100.00
Miscellaneous expenditure	400.00
Total	2,000.00
Annual Income : Fish production	
Assuming average growth of a fish to be 1 kg and survival at 75%, the yield would be = 2000 x 75 = 1500 kg/0.2 ha	6,000.00
Annual income through fish (at the rate of Rs. 4/kg fish)	
Annual expenditure	2,000.00
Net income/0.2 ha/annum	4,000.00
So net income/ha/annum	20,000.00

Fish Culture in Sewage-fed Ponds

As the sewage water is very rich in phytoplankton production, the stocking density of fish fingerlings can be 10,000 Nos/ha against 5,000 Nos/ha of fish farms fed with freshwater. Furthermore, there is no need to add inorganic fertilisers in sewage fish ponds; such fertilisers are essential for freshwater fish farms.

Stocking ratio given in Table 1 could be followed in sewage-fed fish ponds.

Air-breathing and low oxygen tolerating fishes like murrels can also be grown together with those mentioned above: however, they should be introduced only after the fingerlings of other fishes have grown to a considerable size. Otherwise murrels may prey on other fish fingerlings. But in oxidation ponds murrels can be cultured with the fast proliferating trash fish namely *Tilapia mossambica* as the latter forms food for murrel fingerlings.

Fish production from sewage systems has been found to vary from 7.5 tonnes to 8.5 tonnes/ha/annum depending upon the survival rate of fingerlings which may be from 75 to 85%.

4

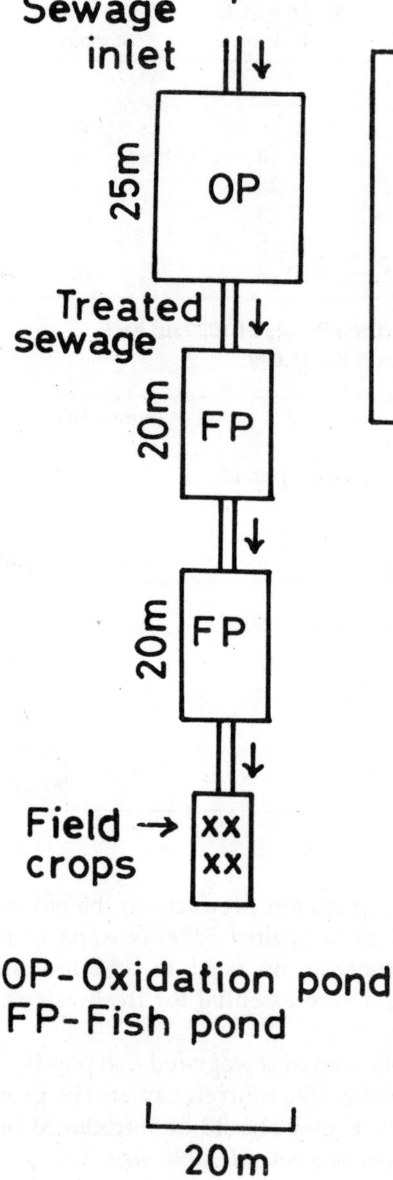
Sewage
inlet
25m
OP
Treated
sewage
20m
FP
20m
FP
Field
crops
xx
xx
OP-Oxidation pond
FP-Fish pond
20m

5

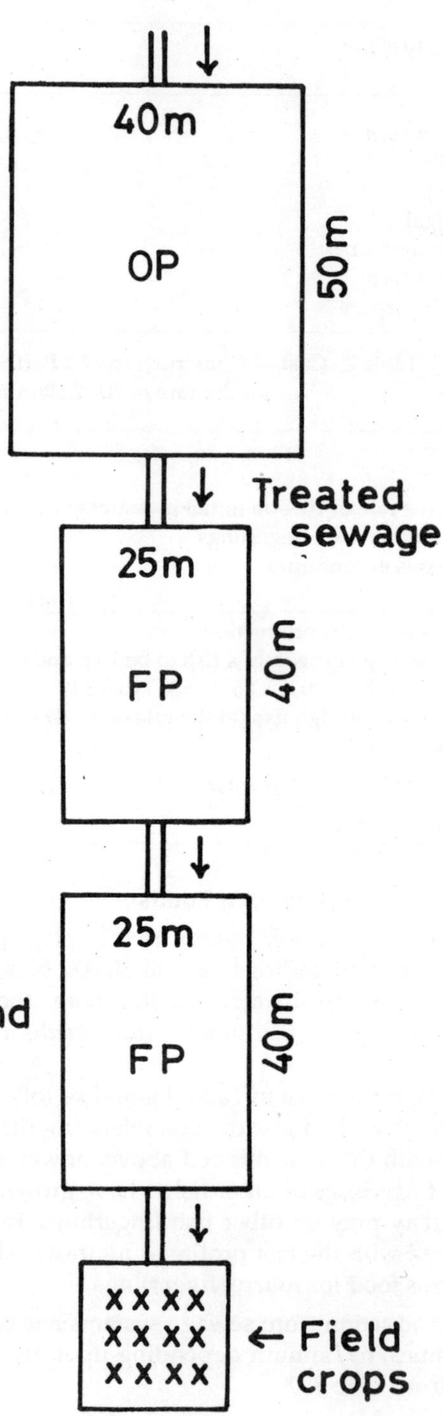
40m
OP
50m
Treated
sewage
25m
FP
40m
25m
FP
40m
xxxx
xxxx
xxxx
Field
crops

Quality of Fish Harvested from Sewage Farms

Bacterial tests were conducted by scientists of the Department of Fishery Science, Tamil Nadu Agricultural University, Coimbatore to examine bacterial contents of the skin, gut and muscle of fishes grown in sewage and freshwater. From experiments it was evident that species of *Salmonella* were not present in any of the fishes. *Shigella flexneri* was isolated from fishes reared in sewage-fed ponds. But neither of them nor the pathogens were present when the fish were steamed, fried or curried. Furthermore, preparations of fish harvested from sewage were found to be tastier than those of freshwater.

Sewage Water for Other Crops

The excess sewage water drained from the sewage-fed fish ponds can be used for irrigation purposes, especially in the cultivation of fodder crops, vegetable crops, coconut, palms, etc. Income can be increased considerably.

Model Plans

Depending upon the amount of sewage discharge, fish ponds can be installed. The plan of sewage-fed fish culture unit for 500 residents suited to educational institutions is shown in Fig. 4 and the unit suitable for local bodies is shown in Fig. 5. A model of sewage-fed fish culture unit and the role of its different components are shown in Fig.6.

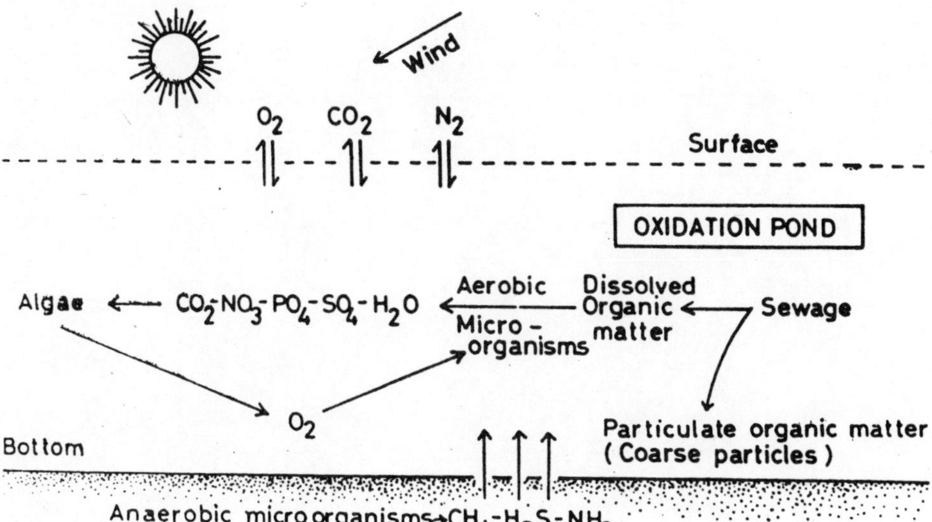

Economics of Sewage-fed Fish Culture

The expenditure on sewage-fed fish culture is meagre in comparison to freshwater fish culture. Expenditure on the latter is mainly due to the formation of fish ponds and sewage channels. Table 2 gives the economics of sewage-fed fish culture. The cost of oxidation pond is not included as it is to be erected by the civic body concerned. (If murrels are cultured in oxidation ponds and the excess sewage water is utilised for the cultivation of various crops the revenue could be further enhanced).

Organisations like town panchayats, municipalities, educational institutions and housing boards who plan out drainage systems can follow the sewage-fed fish culture techniques and increase their income by producing fish. If such ventures are undertaken by farmers and public, they will obtain not only 'wealth from waste' but will also keep the environment clean.

5

Fish Farming Using Sewage Wastes

Fish are known to convert biowastes into edible protein efficiently. Human wastes such as sewage, septic tank sludge and night soil can be profitably recycled for fish culture. As these wastes replace the expensive supplementary feeding in fish culture systems, the cost of fish production through human wastes is very less.

Further, owing to the attenuation of pathogens in fish ponds, the pollution hazards and the prevalence of insanitary diseases could be controlled to a greater extent. Hence, it is imperative to expand the human waste-fed fish culture practices by civic bodies, educational institutions, etc.

Sewage-fed Fish Culture: The domestic sewage has poor dissolved oxygen content and it should be treated prior to use in fish culture through stabilisation in oxidation ponds. In these ponds, the sewage is first held for about a week when its coarse particles settle and the dissolved organic matter is decomposed by aerobic microorganisms resulting in the formation of inorganic nutrients such as nitrate, phosphate, sulphate, etc.

The particulate matter settled at the bottom, on the other hand, is partially decomposed into various organic acids, methane, hydrogen sulphide, ammonia, etc. owing to the limited availability of oxygen. As a consequence of the aerobic decomposition, the nutrient status of the water is improved leading to proliferation of algal cells and increase in the dissolved oxygen level which help in the further oxidation of the partially decomposed organic substances into inorganic substances.

As a result of these the sewage in the oxidation ponds is clarified with rich oxygen and nutrients and is made suitable for fish culture.

The oxidation pond is made of concrete and sewage is transported to them by slopy open channels. Provision is also made to discharge the sewage at the bottom of the oxidation pond, so that, there will be least disturbance in the supernatant portion which is to be taken by gravitation to the adjacent series of earthen fish ponds. The number of fish ponds should be such that their total area is equal to half the area of oxidation pond.

Recycling of Septic Tank Sludge: The reluctance of people to consume the septic tank sludge-grown fish may be overcome by stabilising the sludge in oxidation pond with the addition of freshwater at 1:3. In the latter, the nutrients recycled from organic waste help in the development of fish food organisms viz. plankton which is an ideal protein-rich food source for culturable fish.

Courtesy : The Hindu

Alternatively, culturing of tilapia and other trash fish can be undertaken to make them serve as a direct live food source for carnivorous murrels and catfish or as one of the chief ingredients in the preparation of fish feed for use in carp culture.

Recycling of Night-soil: As human excreta contain about 30 per cent nitrogen, 5 per cent phosphorus and 6 per cent potassium, recycling of such wastes for use in fish culture would be profitable. In China, Taiwan, Thailand, Malaysia and Indonesia, fish ponds constructed below the toilets serve as stabilisation ponds. Human excreta added at 200t/ha/yr are known to increase production of food organisms which in turn help maximise fish production.

A more or less complete destruction of pathogens, however, could be accomplished, if the human wastes were treated either in a biogas unit or by composting before being added to fish pond.

In the ground surface continuous aerobic composting method, which involves night-soil and vegetable matter, the increased temperature formed, kills pathogens present in the night-soil and the latter thus is made suitable as a fertiliser for fish pond.

Fish yield: Freshwater carps such as catla, rohu, mrigal, commoncarp and silvercarp are suitable for human waste-fed fish culture. Considering the rich nutrient status of human wastes, a fish yield of 6t/ha can be obtained by stocking fingerlings at 20,000/ha. The revenue through human waste-fed fish culture can be further augmented, if the excess treated waste is used for irrigation of fodder, crops, vegetable crops, coconut, palms, etc.

To ensure greater safety of fish flesh free of pathogens, the waste-grown and harvested fish may be maintained for a week in clean water which may eliminate the disease causing organisms that survived the earlier treatments. Proper cooking of the fish further renders the fish completely safe for human consumption.

6

Freshwater Prawn-cum-Carp Farming — A New Polyculture Practice

Though a lot of work has been done on composite fish culture incorporating Indian major carps and exotic carps having different feeding habits and considerable production achieved, no large scale monoculture of freshwater prawn or polyculture of freshwater prawns and carps has been attempted in our country. Jhingran (1983) states 'among the increasing number of aquatic animals that have been coming to the fore as candidate species for aquaculture, - the freshwater prawns are perhaps the most recent'. Keeping this in mind, culture of surface and column feeding carps and bottom feeding freshwater prawns could be undertaken as a polyculture practice in our Indian freshwater ponds to maximise yield.

In India 34 species of *Macrobrachium* comprising 20 freshwater species and 14 coastal water species have been reported. However, only 2 species viz., *M. rosenbergii* and *M.malcolmsonii* have been found to be viable for monoculture and polyculture practices owing to their large size, fast growth and shorter life history than that of penaeid prawns.

Identifying Features of *Macrobrachium* spp.

M.rosenbergii, otherwise called 'giant freshwater prawn' (Fig. 7) has dark grey and longitudinal streaks of darker and lighter colour. The largest pereopod of adult male is bright or dark blue or lavender and orange at joints. The eggs of berried females are yellow. This species is identified by a long and slender rostrum with 11-14 teeth on its dorsal margin and 8-14 teeth on the ventral margin. Both the sexes grow up to 35 cm (male) and 24 cm (female) and attain a weight of 200 g. Though this species is found distributed mainly in estuarine areas and rivers, it is also sometimes found at sea.

M.malcolmsonii (Fig. 8) otherwise called 'monsoon river prawn' is darker grey dorsally and paler ventrally. It is identified by a rather long rostrum with 7-11 teeth on the dorsal margin and 4-7 teeth on the ventral margin. The male and female grow up to 25 and 20 cm respectively.

This species is mostly found distributed in rivers and estuarine areas.

Food and Feeding Habits

These bottom feeders are omnivorous and feed on aquatic worms, insects, small crustaceans, molluscs, foraminiferans, nematodes, offals of fish and other animals, vergetable matter, filamentous algae, tender leaves and stems of aquatic plants. Bhimachar (1965) stated that food preference by species of *Macrobrachium* is almost non-existant. However, there exists food preference in relation to age groups in that the juveniles resort to feeding mainly on diatoms.

Courtesy : Fishing Chimes

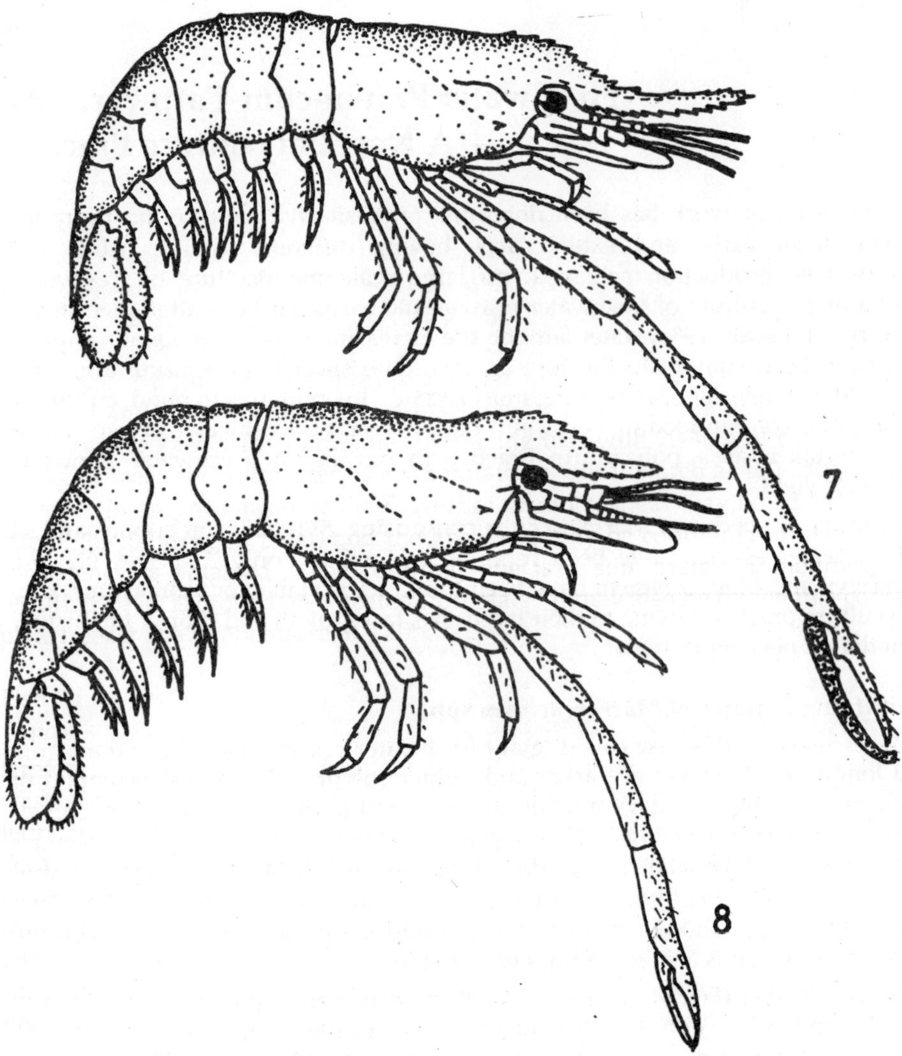

Age and Growth

In *M.rosenbergii,* males grow faster than females and matured male and female moult six and five times respectively in a year.

Reproduction

During the breeding, the males predominantly stay in freshwater zone while the females move to the brackishwater areas and thereby the sex ratio is altered. The females predominate in monsoon months and male in winter. However, both males and females show equal distribution in summer. Invariably these species mature at the end of first year. The size at first maturity for

M.rosenbergii is about 14 cm and for *M.malcolmsonii*, 4cm. The fecundity increases with the size and the fecundity rates were found to be 7000 to 500,000 for *M.rosenbergii* and 4000 to 60000 for *M.malcolmsonii*.

The breeding season of freshwater prawns coincides with monsoon. The breeding period in east coast for *M.rosenbergii* is December-July with a peak period of March-May and in west coast, the corresponding periods are August-December and October-November. *M. Melcolmsonii* breeds in east coast during April-November with a peak period of June-July. Temperature and salinity are found to influence the breeding migration of freshwater prawns. The larvae (2mm) hatch out in brackishwater and they require a salinity of 10-25‰ for better survival. In two weeks, the young larvae become postlarvae which after a sojourn of a week or two in brackishwaters go back to the rivers.

Culture Practice

The ideal size of the production ponds for prawn culture is 0.2 ha. The optimal depth required is 0.7-1.0 m.

The slope of wet side bunds may be 1:3 and of dry side, 1:2. Depending on the nature of the pond bottom, lime should be administered. Quick lime may be applied at the rate of 1000 kg/ha. The water usable for the production ponds should have a pH of 7-8.5. If the pH of the water goes above 8.5, the same may be stagnated in the production ponds for about 2-4 weeks prior to stocking with prawn seed.

As prawns feed mainly on detritus, production ponds intended for monoculture of prawns need not be fertilised. However, for growing prawns and carps together, the ponds need to be fertilised as in composite fish culture ponds. The ponds are first fertilised with organic manure (cowdung) at the rate of 10-20 t/ha. It would be better if this manure (one-sixth of the total quantity) is dissolved and added in the pond water 15 days before the release of fish and prawn seed. The rest of the manure is added monthly in equal instalments. The other chemical fertilisers to be added are the ammonium sulphate, urea, superphosphate and muriate of potash at the rate of 450, 200, 250 and 40kg/ha respectively and are added every month in equal instalments.

Stocking of Prawn and Fish Seed

Four week old juvenlle prawns (3-5 cm) and carp fingerlings could be stocked at the rate of 50000 and 5000/ha respectively. While selecting carps, it is better if bottom feeding common carp and mrigal are avoided. Catla, rohu, silver carp and grass carp may be stocked in the ratio of 2:1:2:1. As the growing prawns may have a strong territorial instinct which often lead to the death of smaller ones, hideouts such as damaged nets or palm leaves may be installed at various places in the production ponds. These hideouts also harbour organisms which in turn serve as good food source to the growing prawns.

Supplementary Feeding

For growing prawns, supplementary feeds made of trash fish, beef liver, pila (mollusc), tapioca flour and pig dung may be given once in three days at the rate of 5-10% body weight. Feeds containing 25-35% and 40% protein have been found to give better growth in *M.rosenbergii* and carps respectively. For carps particularly during the periods of absence of live food (plankton) in pond, food balls of groundnut oilcake powder and rice bran mixed in the ratio of 1:1 may be given.

Production

As these prawns attain marketable size in about five months, two crops of prawns could be had. *M.rosenbergii* has been found to show a maximum growth of 100g (20cm) in a period of nine months. In pond monoculture, yields of 2.6 t/ ha/yr and 4.7 t/ha 110 days have been reported for *M.rosenbergii* under the stocking densities of $5/m^2$ and $83/m^2$ respectively.

It is also worthy of mention here that *M.rosenbergii* could be cultured alongwith milkfish and mullets in brackishwater ponds having a salinity range of 12-25% and an individual growth of 100 g/5 months has been reported under a salinity of 7.2% and a stocking density of 29400-166600 nos/ha.

7

Tilapia-Tarpon Culture in Fresh and Brackish Waters

Tilapia (*Oreochromis mossambicus*), (Fig. 9) an exotic freshwater fish has been found widely distributed in the majority of the inland water systems in the country. However, its culture along with Indian major carps (catla, rohu and mrigal) and chinese carps (silver carp and grass carp) was not favoured in the ponds due to its early maturity and unwanted proliferation which leads to not only overcrowding and stunted growth but also competition with the aforesaid quality carps.

9

The peculiar characteristics of tilapia such as its tolerance to wide variations in water temperatures (8 to 42° C), salinity (up to 7.5 per cent) and pH (5-11) and low oxygen level (as low as 0.1 ppm); production of a large number of young ones all through the year and feeding on a wide variety of plants and animals make this species now suitable for culturing in fresh and brackishwaters with tarpon (*Megalops cyprinoides*), a marine carnivorous fish having wide tolerance in varying salinities (Fig. 10).

Courtesy : The Hindu

10

Seed Fish

Tilapia normally spawns in its third month in water temperatures of 20-30°C and salinity of less than 3 per cent. In a year, six to 11 spawnings with an interval of about three weeks in between have been recorded. However, the differentiation of sex is possible only when they gain a weight of 50 gm.

The male is generally larger than the female and possesses a dark body and deep-red fins. The female has two orifices in the genitalia as against one in the male. Before spawning, the parent fishes using the mouth make a saucer shaped depression (30-90 cm long and 6 cm deep).

During spawning, a single female lays about 1000 eggs which are fertilized by the male with its sperm. The fertilized eggs—yellowish, elliptical and about 2.5 mm in diameter—are immediately picked up by the female in its mouth. The eggs are rolled around and are hatched into larvae in 60 hours at 28-30°C. The larvae swim out of the mouth of female fish after a week.

The tarpon reaches maturity when it attains a weight of about 400 gm and spawns twice a year. When it migrates to estuaries and rivers in April-October, its seed could be collected.

As the fry of tilapia and tarpon feed mainly on zooplankton such as cyclops, daphnids, cyprids and rotifers, the nurseries have to be fertilised with cowdung for the production of these organisms.

Production Pond

The pond is also fertilised with chemical fertilisers (urea 75 kg, super phosphate 100 kg and muriate of potash 15 kg per ha) and cowdung (4000 kg/ha) for the development of zooplankton as a chief food source for tilapia and tarpon fingerlings.

For better results, these fertilisers are applied alternately once in 15 days for a period of about four months or until the tilapia attain a weight of about 200 gm each and produce their young for the growing tarpon as a chief food source.

The fingerlings of tilapia and tarpon are stocked at the rate of 30,000 and 3,000 per ha, respectively. Napier grass or tender weeds added daily at 50 per cent of its body weight would serve as food of adult tilapia in both fresh and decayed states.

Harvest

To avoid the overcrowding of tilapia, about half of them (about 3 tonnes of it with an average individual weight of 200 gm may be harvested as first crop after four months and marketed. Depending on the availability of water in the production pond, the rest of tilapia and tarpon may be harvested after six or eight months.

A second crop of tilapia (7.5 tonnes) and a tarpon yield of 1.5 tonnes would be possible. The tarpon fed daily with rice bran and groundnut oilcake at 1:1 in 2 per cent body weight ration had grown to 600 gm each in a period of six months in the freshwater pond of the Fisheries College.

In the absence of tarpon fingerlings, farmers can also undertake culture of male tilapia. For this, male tilapia are segregated and grown in the production ponds at the rate of 50,000/ha. Male tilapia may also be obtained through sex reversal by rearing the fry of unknown sex for a period of about two months with feeds containing 40 per cent rice bran, 30 per cent tapioca flour, 30 per cent yeast and methyl or ethyl testosterone added at 30 mg per kg feed.

Females fed four times a day at 2-5 per cent of their body weight are also known to change their sex within two months.

8

Milkfish-Prawn Farming in Brackishwaters

Brackishwater, a mixture of seawater and freshwater has a salt content of 0.3 per cent and is highly fertile due to the influence of the inshorewaters rich in nutrients and fish-food organisms through tidal waves from the sea. The young ones of fish and prawn migrate to these areas which serve as good nursery grounds.

Therefore, areas adjacent to brackishwaters can profitably be utilised for forming ponds to culture milkfish (*Chanos chanos*) and prawns such as *Penaeus indicus* and *Penaeus monodon* (Fig. 11-13). This type of culture of fish and shellfish otherwise called as 'polyculture' would be a promising venture.

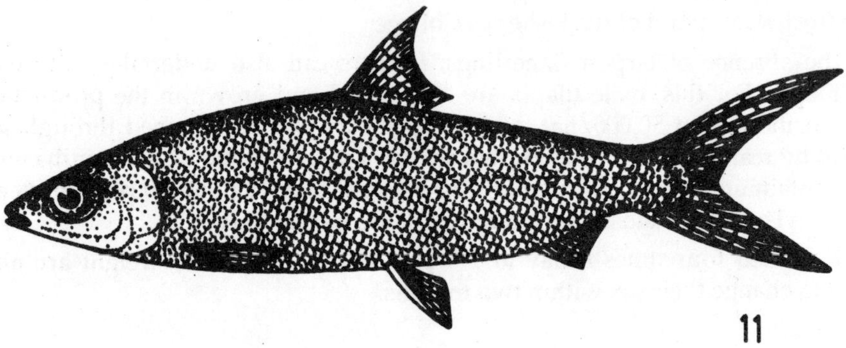

11

Milkfish fry of about 2 cm length generally swarm in large shoals in shallow tidal creeks having calm and clear water particularly during April-June and October-December. They may be collected using scoop or drag nets. Seed prawns are invariably available in our coast round the year. However, during spring tides (new or full Moon days) these could be collected in large numbers.

Courtesy : The Hindu

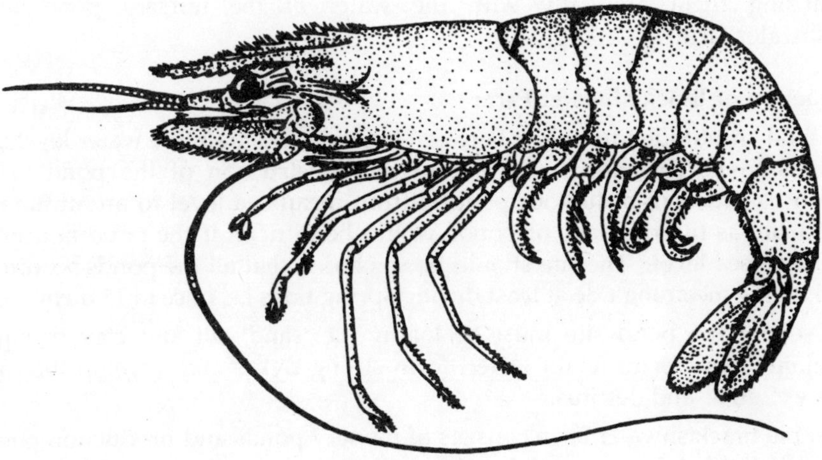

12

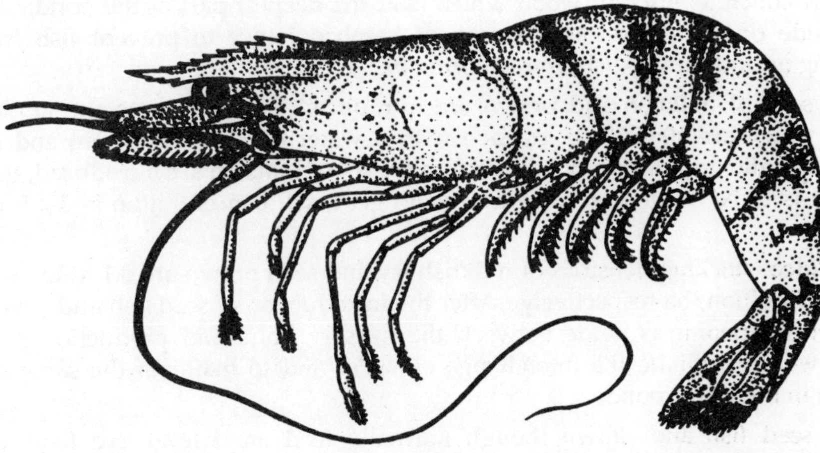

13

The milkfish fry collected from brackishwater areas may be introduced in the nursery ponds after acclimatising them in cement tanks containing tapwater and salt adjusted to a salt content of 0.5—1.2 per cent. Each tank of 3 sq. m area and 20 cm water depth may hold 30,000 fry. The fry are fed twice a day with powdered groundnut oilcake wheat flour, etc. Water is changed once or more daily. The seed prawn may be introduced into the nursery pond after acclimatising them gradually with the water of the nursery pond in a brackishwater farm.

Construction of Brackishwater Farm

The tidal amplitude (difference between low tide and high tide water levels) of the site is to be considered first before the construction of the pond. The optimum elevation of a dug-out pond is from mean sea level to an altitude of about 50 cm, as full drainage of ponds would be difficult if the pond bottom is below mean sea level. The site should be so chosen that all the ponds would be flushed by the incoming tide at least during spring tides i.e. once in 15 days.

The soil of the pond site must be loamy (i.e. sand, silt and clay in equal proportions) in order to retain water, form strong dykes and develop the food sources viz. algae and detritus.

A simple brackishwater farm consists of nursery ponds and production ponds which are divided by inner dykes having slopes ranging from 1:1.5 to 1:2. Depending on the land size, 1 per cent of it may be used for the formation of nursery ponds and the rest for production ponds.

The water for the farm is mainly by gravitational flow through a main channel linked with sea. Both the supply and drainage of water are ensured by a single sluice of concrete, brick or wood which is at the deepest part of the pond. The inner side of the sluice has a screen of bamboo lattice to prevent fish from escaping or undesirable fish from entering in.

The seed of milkfish and prawn are introduced separately into the nursery ponds which are 100-200 sq. m each with a water depth of 20-50 cm and are located parallel to stocking ponds. Before the fry of milkfish are introduced, their nursery is to be diluted with freshwater to get a salt concentration of 1-1.5 per cent.

The ideal stocking densities of milkfish fry and seed prawn are 0.1 million/ha and 1-1.5 million/ha respectively. After the introduction of seed fish and prawn, a screened opening is made between the nursery pond and production pond which would facilitate the interchange of water and to maintain the same salt content in these two ponds.

The seed fish and prawn though normally feed on natural live food viz. plankton, they may also be fed with groundnut oilcake powder. The nursing may be done for a week and the screen between the nursery and production pond may than be removed. Now the juveniles would swim into the production pond of their own accord.

Management of Production Pond

In order to develop the natural food sources for the growing milkfish and prawn, the bottom of the rearing pond is first exposed to Sun until the soil cracks and crumbles easily. Then it is fertilised with any of the organic manures such as chicken droppings, cowdung or pig manure at the rate of 500 kg/ha.

Seawater is let into a depth of 10-15 cm and is allowed to evaporate until the bottom is dry. If there is no formation of algal bed on the bottom, the latter is again fertilised and then the seawater is let into the pond as a second time to a depth of 10-15 cm. At this time, it is advisable to cast mahua oilcake or tea seed cake at various places in the pond in order to eradicate undesired animals if any.

The acclimatised seed fish and prawn are stocked in the pond at the rate of 5,000 and 40,000 numbers/ha respectively.

During rainy season, if the salt content of pond water goes too low to support good growth of algae, the growing fish and prawn may be fed daily with food balls prepared by powdered groundnut oilcake (20 kg/ha) and rice bran (30 kg/ha). Organic manures may also be applied as and when required to maintain the growth of algae.

As the milkfish and prawns reach marketable size in about 8-10 months and three-four months respectively, one crop of milkfish and three crops of prawns may be raised in a year. Following intensive management practices, a production of 1000 kg milkfish and 400 kg prawns/ha would be possible.

9

Crab Culture in Coastal Ponds

Marine crabs form one of the important protein-rich food sources easily available almost throughout the year. Among the edible species, *Scylla serrata*, (Fig.14) commonly called mud crab or green crab is known for its delicacy, high price and export value. As it is the only fast growing and larger sized (2-3 kg) species, it is more suitable for culture in areas adjoining the productive and seed-rich brackish water and mangroves of the country.

14

As the crabs bore holes along the dykes and cause leakage, under-developed coastal ponds with numerous mounds and unsubmerged regions are more suitable for crab culture. To prevent the escape of crabs by climbing the innerside of the ponds are fenced with bamboo matting to a height of about one metre.

Production of Seed

The mud crabs having a carapace width of more than 11 cm, mature in a period of 10 months. Before actual mating, the male and female pair for about four days and the female moults. The copulation then takes pace for 7-12 hrs. The sperms are retained by the female and fertilization takes place after weeks or months.

Courtesy : The Hindu

A single copulation would fertilize the different batches of eggs which amount to about two million. The fertilized eggs attach themselves to the females which migrate to the off-shore for spawning. The eggs hatch out into first zoea in a week or two.

By a series of moults, five planktonic zoeal stages and one megalopa stage are derived in a month. While the zoeae feed on phytoplankton, the megalopae being predators, feed on invertebrate and fish larvae. The megalopae reach the bottom and metamorphose into juvenile crabs which migrate along with the tidal waters to the plankton-rich nursery grounds viz. the inshore waters and estuaries.

In the sea water-fed coastal ponds, though mud crabs normally mature and spawn, the embryonic development and successful hatching would not take place owing to their low and varying salinities. Further, the abundance of mud crab seed is seasonal and localised in the brackishwaters. Hence, the females are subjected to eye stalk ablation to get rear-round production of berried crabs for the supply of larvae.

Eye Stalk Ablation

In prawns and crabs, the 'X organ sinus gland' of the eye stalks inhibits egg production through its 'gonad inhibiting hormone' (GIH) and the brain and thoracic ganglia on the other hand promote egg production through their 'gonad stimulating hormone' (GSH). Under favourable environmental conditions, the secretion of GIH is however suppressed.

Based on this principle, precocious maturation of ovaries and moulting are obtained through the removal of one of the eye stalks in larger females by electro-cauterisation, in which one of the eye stalks is cut at the base with a pair of scissors followed by sealing with a pencil type of soldering iron. Eye stalk ablated female and unablated male brood stock are well maintained in laboratory with protein-rich diets before allowing them to mate at the appropriate time. After the larvae are hatched out, they are fed with milk powder and egg yolk under 24-hour illumination which induces the larval feeding.

Stocking and Feeding

Juvenile crabs having a carapace width of 7-10 cm are stocked at 10,000-30,000 per ha.

To make the operation more profitable, females that have just had copulation may be reared for one or two months until their full gonad development. However, as post copulation females may not be available in required quantities, small ponds may be maintained to hold different batches of crabs which are collected at intervals and grown.

Though the growing crabs prefer to crawl at the bottom for feeding on algae, decaying animals and detritus, they may also be fed with trash fish like tilapia, fish offal and visceras, slaughter house waste and clam and snail meat at 5-10 per cent of their body weight to enhance the growth. The meat of aquatic snails is known to help in the early maturation of crabs. The supplementary feeding is done daily in the evening as the crabs feed normally after sunset.

Growth and Harvest

The mud crabs can be harvested within six months when they grow to marketable size (more than 12 cm and 220 gm). Under intensive feeding and judicious management, an yield of 900 kg/ha can be easily obtained. Marketable crabs are captured by hand with the help of a small dip net or with large dip net baited with trash fish.

Male crabs if copulated several times would have little meat and fetch low price. Hence they are harvested before that stage. The harvested crabs may be transported live in wet gunny bags or heavy straw ropes.

In many countries *S. serrata* is cultured mainly for its meat-rich chelate legs which are amputated periodically for marketing. Mud crabs can also be cultured along with milkfish-prawns in modified coastal ponds. Similarly, in the pokkali rice fields of Kerala and in the Bheries of West Bengal, *S. serrata* can be included along with fish and prawns after the harvest of the first rice crop.

10

Oyster Farming in Brackishwaters

Edible oysters were not much relished in India until a few decades ago. But, of late, the farming of oysters is gaining momentum as a source of cheap protein-rich food (protein 10 per cent, fat 2 per cent, carbohydrate 6 per cent. Vitamins A, B and B12 besides nicotinic acid and minerals).

Oyster farming on a small-scale has been attempted in some of the coastal areas of India, but brackishwater oyster farming is hitherto unknown. In a country with about 2 million ha of fertile brackishwaters, oyster farming could be undertaken as a commercial venture. Of the four species of oyster viz. *Crassostrea madrasensis*, *C. cucullata*, *C. gryphoides* and *C. discoidea*, occurring all along the Indian coasts, *C. madrasensis* is more suitable for brackishwater farming. (Fig. 15).

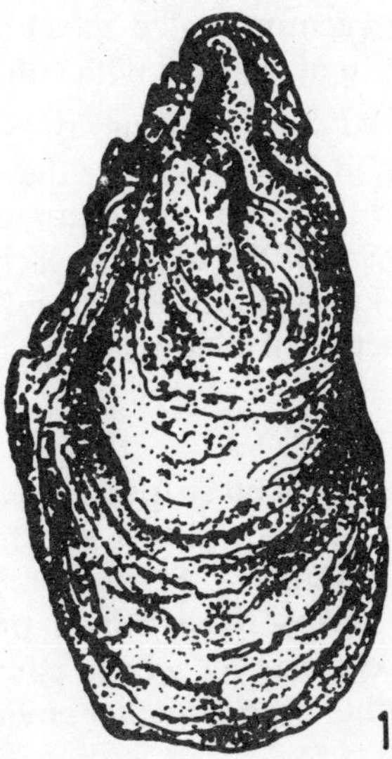

15

For on bottom culture, firm inter-tidal and sub-tidal areas free from excessive winds, wave actions and predators like boring molluscs and crabs are ideal. To ensure adequate spat (seed oyster) supply, areas of extensive oyster beds should be available adjoining the site. Tidal and current flow must be sufficient to change the water and bring more natural phytoplankton food for growing oysters in the off bottom culture. Above all, the salinity of the site should be 20-30%.

Courtesy : The Hindu

Collection of Spat

The spat for farming may be collected naturally from oyster beds or through suspending materials like lime coated coconut shells, oyster shells, corrugated asbestos strips, tiles and nylon fibres (spat collectors) below the rafts where the adult oysters are grown or distributed naturally.

The first phase of oyster farming is 'seedling production' and the second phase is 'rearing' during which, the young oysters are fattened for market. The first involves sowing seed oysters directly on firm intertidal bottom.

A rack suitable for shallow waters consists of vertical casuarina poles driven to the bottom which support rows of horizontal poles. The trays (90 cms x 60 cm x 15 cm) are made of 1 cm meshed nylon webbings supported on all sides with 2 mm GI rods. The spat obtained naturally or from collectors are held in trays suspended from the horizontal poles of the racks and grown for about five months.

A raft (squarish or rectangular) suitable for fairly deeper waters is made of bamboo or casuarina poles 10-15m long which are laid parallel to each other about 0.5m apart and fastened by nylon ropes or wire lashings to lateral beams. Floating of rafts is aided by vacuum sealed metal drums, the number of which varies depending on the weight of the rafts with culture materials.

The rafts are also anchored from drifting away by wind and waves. The young oysters (3-5cm) grown in rack's trays are then transferred to another set of trays hung from the rafts and the culture is continued for about one year. In the trays of the rafts, 80-90 per cent of the oysters reach a marketable size of 15 cm with a meat weight of 15-20 kg each. Considering 7-10 per cent of meat content of oysters, a total meat production of 18 tons/ha/yr is possible.

Stick Method

Here, cheap wooden sticks, bamboo strips or tree branches are driven into the bottom of the shallow brackishwaters. These substrata are ideal for the natural settlement and growth of oysters. Fertile brackishwaters near Tuticorin experimented with bamboo strips have shown promising results (calculated production of 15 tons of meat/ha/yr) which pave ways for the commercial farming of edible oysters in other brackishwater environments.

After harvesting, the live oysters, owing to their filter feeding habit and possible accumulation of pollutants should be left in cement tanks containing fresh sea water for two days followed by treating them for about one hour in freshly chlorinated (3 ppm) water. By these processes, the oysters may get rid of harmful bacteria, if any, by evacuating their stomach contents and become suitable for consumption.

The shelled oysters may be despatched to nearby places in gunny bags. Fresh oysters meat may also be consumed with pepper and lime juice.

11

Pole Farming of Edible Oysters in Brackishwaters

In India 1.4 million ha of fertile and pollutant-free brackishwaters are available all along the east and west coasts. These areas serve as feeding and nursery grounds for many species of fish and prawns. However, at present, only about 30,000 ha of brackishwaters find application in aquaculture. These shallow areas offer vast scope for the farming of edible oysters belonging to the species, *Crassostrea madrasensis.*

Farming Techniques

Though a number of oyster farming practices suitable for inshorewaters and deeper open seas are available, methods for shallow brackishwaters need to be standardised. In this context, the 'pole technique' experimented for the first time in our country has yielded promising results. In countries like France, this technique, otherwise called 'Bouchot culture', is commonly followed for large-scale farming of edible oysters. However, it is worth mentioning that while the harvestable size of oysters is obtained in 3-4 years in temperate seas, it is within 10-12 months in tropical seas. This factor, in combination with the fertility of brackishwaters, help in the faster growth of edible oysters. In the pole technique, many rows of cheaper casuarina or bamboo poles of 2m high are installed in shallow brackishwater areas for the settlement and growth of oyster spats.

The spawning of oysters though occurs year-round in tropical seas, peaks are observed invariably during the end of dry season or the onset of rains. The planktonic larvae which emerge after successful spawning and fertilization settle on suitable substrata in 2-3 weeks. This part of the life cycle of oysters infact facilitates the pole culture to a large extent. By monitoring the spawning periodicity of oysters and installing poles at right time, rich settlement of oyster spats in the poles can be expected.

Growth and Yield

As the pole - settled oyster spats feed chiefly on nano-phytoplankters which are generally abundant in brackishwater areas, the growth of oyster in poles (Fig. 16) and bamboo pens (Fig. 17) installed here was found to be more than that of oysters grown through racktray method in the nearshore waters. For instance, in the nearshore waters, the annual maximum growth of oysters was 11 cm against 13 cm and 17 cm in poles and pens respectively. The estimated values of oyster yield (shell-on weight) in the inshore waters and brackishwaters were 180t (rack-tray), 325t (pole) and 350t (pen) per ha. By considering 7 per cent as the average

Courtesy : Aquaworld

meat content of oysters, the meat yields were 12.6t, 15t, and 17.5t/ha respectively. This study, the first of its kind in brackishwaters, as so far edible oyster culture has been attempted only in coastal waters of moderate salinities, paves way for the commercial farming of edible oysters in association with finfish farming in brackishwater pens or poles.

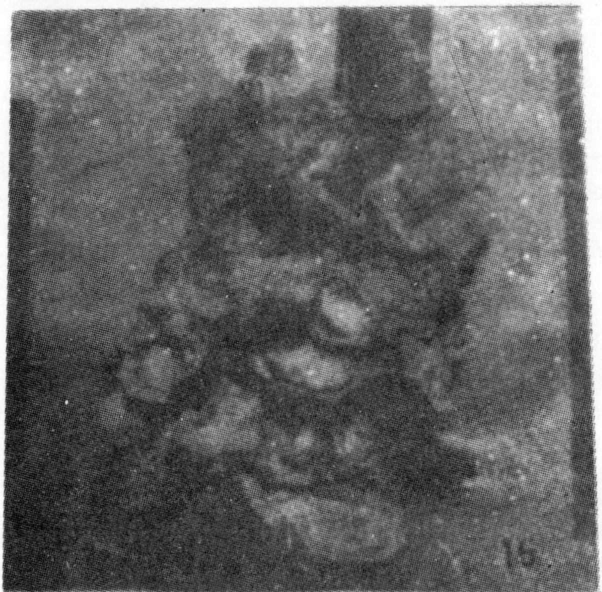

17

Oyster in Human Nutrition

The succulent meat of oyster has reputed powers ranging from aphrodisiac to restorative. Besides easy digestibility, it has moderate protein content (10 per cent) and is a valuable source of essential vitamins (vitamin A, B and B12), aminoacid (Nicotinic acid) and minerals (calcium, phosphorus, iron, sodium and potassium). The oyster meat after proper purification may be consumed raw with pepper and lime. Alternatively, it may be pickled or brined and canned. Steaming of meat followed by smoking is popular in Western countries. Oyster omelette or sandwich is an attractive dish in tropical restaurants. Oyster sauce is yet another popular product, especially in Thailand and Malaysia. It is prepared by boiling oyster meat until a thick broth is produced. The concentrate is then mixed with water and bottled. The resulting meats are sundried and sold for use as flavour enhancers in various dishes.

As oyster farming involves only a small investment and its technology is quite simple, this may be intensified in our hitherto-unutilized brackishwater areas in order to derive the much needed protein for meeting the demand of the ever-increasing human population.

12
Scallop Culture

The scallops, also called as 'Burma shells' are bivalves distinguished from other bivalves by their ribbed and fan-like shape and somewhat symmetrical earlike projections, one on each side anteriorly. Four species of scallops viz. **Pecten plica, P. tranquebaricus, P. crassicostatus** and **P. splendidulus** are found distributed along the Indian coasts. These species are known for their polymorphism (as the individuals possess varying colours and ornamentation). (Fig. 18-22).

Although scallops are found in the depressions of the sea bed, they are abundant in intertidal areas rich in organic matter and planktonic organisms. Naturally-grown scallops are collected from coastal waters of 20-40 m deep, through dredging or diving. The meat of adult scallops which measure about 15.00 cm (shell dia.) is considered to be a delicacy like oysters owing to its high protein (19.5%) and low fat (0.5-1%) content. Although natural and farm grown scallops have a ready market in foreign countries, India is yet to take up its commercial capture and culture activities.

Biology of Scallops

The biology of scallops is interesting as they are hermophrodites carrying both male and female reproductive organs in the same body. They spawn in the spring and autumn seasons. The tiny larvae which emerge after fertilization attach themselves by sticky threads with seaweeds or debris. They remain there until they are large enough to seek suitable substratum on the sea bed. Scallops aged four to five years attain marketable size. They may live up to 12 years. Like other shellfish, older ones yield meat of poor quality than younger ones.

Scallop Culture

Mariculture of scallops can be undertaken on the sea bed in circular or square net enclosures/cages. However, such scallop farms should have good tidal flow and sheltered from wind. A 10 m^2 sea bed would need a net (mesh size 15-20 cm) of 40 m length and 1m height. During installation, the sole rope of the net is weighted with sinkers and its head rope is left buoyant. Casuarina poles or palmyra stems installed at regular intervals would keep the nets in position. Young scallops collected and transported in wet gunny bags are safely penned in net enclosures immediately after netting out all the predators which include chiefly crabs, starfish and whelks. Scallops can also be raised in cages arranged in vertical layers in the shallow nearshore areas. Such practices may give good harvest of the scallops as the natural planktonic food in the column waters is fully utilised.

Courtesy : Fishing Chimes

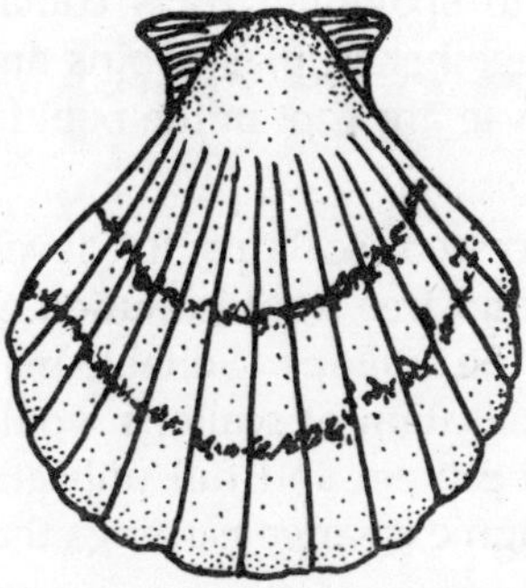

18

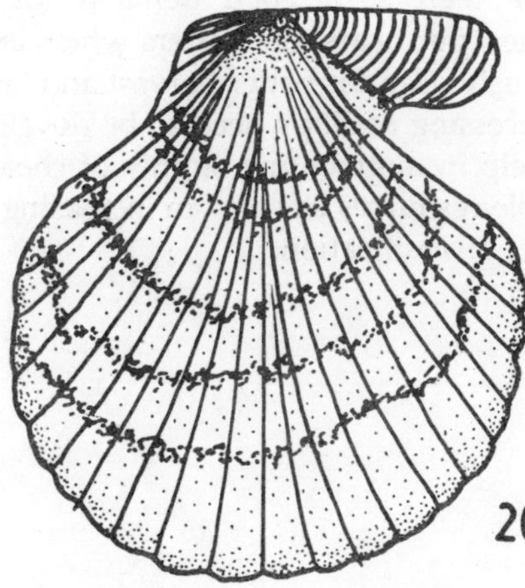

20

19

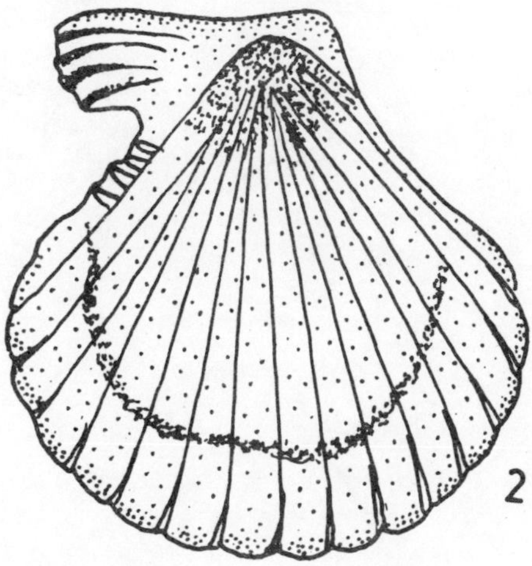

21

22

Seed Production

The marketable scallops are obtained in farms within a period of four months. The seed of scallops for aquaculture can be produced by artificial fertilization employing 'thermal shock' technique. In this process, fully matured scallops are laid out on flaked ice before being transferred to spawning trays containing warm water. The eggs and sperms are mixed together in plastic bins and are then poured into a plastic pool. The emerging larvae are kept in the pool for 3-4 days before being stocked in pens or cages.

Since there is a good demand for the processed scallops in developed countries, our nearshore waters which are fertile and free from pollution can be judiciously used for better harvest and farming of the scallops. Suitable methods for processing scallops need to be developed. Utilisation of scallops would not only help in augmenting supply of cheap animal protein and full utilisation of available resources but also in increasing our foreign exchange earnings through product diversification.

13

Marine Mussels

Mussels occupy a vital place among the molluscan forms. They grow fast and breed profusely. They are excellent organisms for farming in the coastal waters of India. They are low in the food chain and so directly feed on the smaller floating plants namely, phytoplankton, and are able to convert the plant matter into tissue efficiently. They filter water at the rate of 3 litres per hour to collect their food and can tolerate widely varying aquatic conditions. There is no animal that can be farmed in the aquatic environment to excel the production rate of mussels. Hence, they are economically very important.

In India, marine mussels are represented only by a single genus, *Mytilus*. This generic name has been revised as *Perna* and the two species under this genus are *P. indica* and *P. viridis* which are commonly known as brown mussel and green mussel respectively (Fig. 23, 24)

23

Among the green and the brown mussels, the former enjoys a wider distribution along the east and the west coasts of India while the distribution of green mussels is limited to the extreme southwest coast of India, particularly

Courtesy : Science Reporter

from Varkala (near Quilon) to Muttom (near Kanyakumari). Their production (landings) is also low (427 tonnes) while the production of green mussels exceeds 2650 tonnes annually.

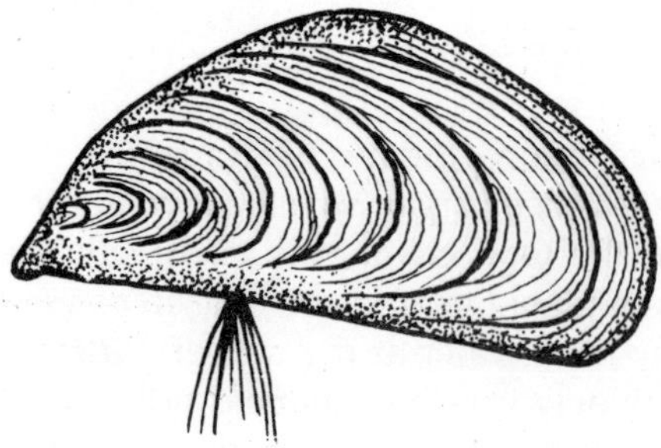

24

Natural Production

In the total marine molluscan production in the world, mussel production contributed by the leading countries is about 3,30,000 tons (shell-on). Annual landings of mussels in India total about 3079 tonnes (shell-on). Mussels harvested from nature are between 40 and 120 mm in size. However, the majority of them are between 75 mm and 100 mm in size. Mussels are gregarious creatures and the natural population density is to the tune of 5 kg per M^2. Such crowding results in slow growth in nature. The growth rate of cultured mussel can be 3-25 times faster than those in the wild. In the case of cultivable *P. indica,* the growth rate is five folds faster. Further, the meat/shell ratio or the percentage edibility is also higher in farmed mussels and their meat is also of good quality.

Methods of Culture

Various methods are available for farming mussels in coastal waters. But in the light of the recent studies carried out in Indian coasts, the raft system of culture has been found to be the most suitable (Fig. 25). In this method, a series of ropes are hung from a raft which floats in the sea. By this technique, it is possible to reap a mussel harvest of 150 tonnes/ha/yr in the case of *P. indica* and 480 tonnes in the case of *P. viridis.* The production rate of mussels has been found to be 15 kg per metre rope in open sea in 5 months; in the bay it is 10-12 kg in 7 months. The growth rate of *P. viridis* is higher than that of *P. indica* and hence *P. viridis* reaches the harvestable size within a period of about 4 months while *P. indica* takes 7-8 months time to attain harvestable size. Through proper planning and management, 2-3 crops of mussels could be harvested in a year.

The growth rate of Indian mussels is higher than those of other countries. *P. viridis* is worth mentioning in this context. It can be reasonably called a short duration species parallel to the short duration crops in agriculture. Taking advantage of these factors mussel farming industry could be developed on a large-scale in India.

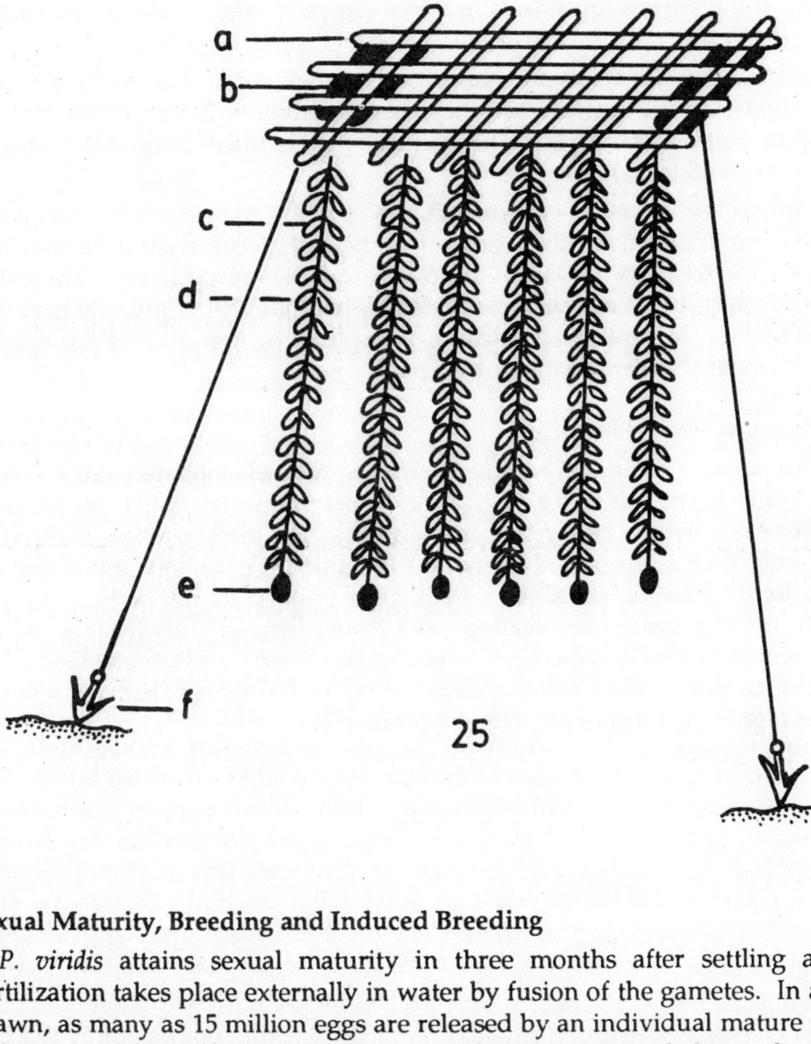

25

Sexual Maturity, Breeding and Induced Breeding

P. viridis attains sexual maturity in three months after settling as spat. Fertilization takes place externally in water by fusion of the gametes. In a single spawn, as many as 15 million eggs are released by an individual mature mussel. Induced breeding of mussels is also possible by vigorous shaking, electric and thermal shock, exposure to chemical solutions and also by pricking the posterior abductor muscle of the mature mussels.

Spat collection: When the mussel larvae come across any hard surface they settle on them and become the spats. Mussel spats are 'seeds' for culture. Materials like roof tiles, cement blocks, asbestos sheets, strings of coconut shells, split bamboo poles, coir ropes, frilled nylon ropes, etc., are used for the attachment of spats on them for later collection. Hence they are called as 'spat collectors'.

Seeding: The collected spats are seeded at the rate of 0.75 kg to 1.4 kg of spats per metre length of culture rope. The optimum size of the spats is usually between 20 mm and 35 mm. They can be transported easily from one place to another. If the population established in the rope is large, it should be suitably thinned later for proper growth of the mussels.

Brown mussels are ready for harvest when they are 50 mm to 60 mm long; green mussels are harvested when they reach a length of 70 mm to 90 mm. The method of harvest is simple. The culture rope is lifted from water and the mussels removed from it.

Nutritive value: Mussel meat which forms 35%-50% of shell-on mussels is soft, tasty and rich in nutritive value. Its protein, fat and glycogen contents vary from 11.08% to 12.6%, 2.38% to 3.02% and 6.36% to 7.91% respectively. The caloric content of the cultured mussel meat is also higher than that of mussels harvested from the wild. Minerals like calcium, phosphorous, iron and iodine are also present in them along with vitamins A and B.

Advantages of Mussel Farming

The advantages of mussel culture are many. Mussels tolerate widely varying salinities and hence are fit for cultivation in brackishwater unlike certain other fish species. Growth in cultured mussels is faster and marketable size is attained in 4 months' time as against 12-13 months in nature. Continuous settlement and availability of spats are possible and 2-3 crops or harvests can be had in a year through proper planning and management. Since they are filter feeders and live mainly on phytoplankton, there is no need to give them supplemental feed which is a must for pond-raised fishes. Yields of up to 480 tonnes/ha/yr is possible which is higher than that of any other organism.

14

Culture of Seaweeds in Brackishwaters– A New and Promising Technology

The seaweeds are sedentary macrophytes growing mostly on the rocks and other plants in the intertidal and subtidal marine environments. Luxuriant growth of the commercially important green, brown and red seaweeds, such as Ulva, **Enteromorpha, Caulerpa, Sargassum, Turbinaria, Gracilaria and Gelidiella,** (Fig. 26-32) exists all along the Indian coastline. As seaweeds find place in human diets, animal feeds, fertilisers, medicines and other industrial products, their exploitation from our coastal localities is increasing day by day. Though, culture of seaweeds has been attempted in the nearshore waters of the few coastal localities of India, no attempt has hitherto been made to cultivate the commercially important seaweeds in the fertile and pollutant-free brackishwaters adjoining many of the estuaries dotting the east and west coasts of India. Fish and prawn culture practices are already in vogue in many of our brackishwater areas. The cultivation of sea-weeds especially **Gracilaria, Gelidiella** and **Caulerpa** which are edible and are known to tolerate wide variations in salinity can well be a viable addition to the species cultured in our brackishwaters.

Culture of Seaweeds in Sheltered Brackishwaters

This involves the floating of webbings supported on all four sides by wooden frames. The webbing consists of ropes of hemp and palmyra fibres or nylon twines which are serving as artificial substrata. Each webbing of 5 x 2.5 m size will have 130 meshes and is installed on poles driven in the intertidal or subtidal area. It is essential that the flat surface of the webbing is close to the water throughout the culture period which is about 3 months. The cuttings (4 cm size) of the matured mother plants which serve as seeds are tied at 100-150g/m^2 to the mesh intersections of the nylon webbings or to the twisted portions of the coir ropes. These fragments show quick regeneration and rapid growth by developing many new shoots from the cut ends. In a cultivation unit of 1-4 ha, 200 webbings may be installed. The harvesting is done by cutting the grown weeds with the help of knives. An yield of 50 t could be obtained from an area of one ha.

Culture of Seaweeds in Brackishwater Ponds

Brackishwater ponds suitable for fish and prawn culture can also be utilised for cultivating several species of seaweeds. Philippines produces large quantities of the green alga **Caulerpa racemosa** in brackishwater ponds every year to meet

---

Courtesy : Fishing Chimes

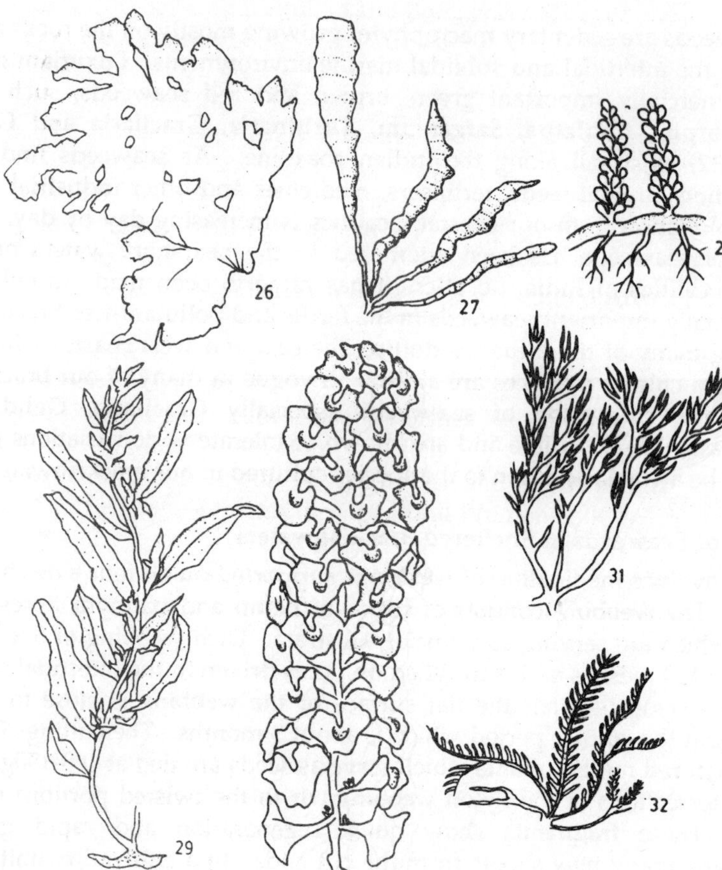
26
27
28
29
30
31
32

the demand in vegetable markets. Brackishwater ponds for seaweed culture should receive tidal waters for exchange at least once in 15 days, i.e. during full-moon and new-moon days. A water depth of 0.5 - 1m and temperatures around 29°C need be maintained besides salinity which should not go beyond 30 p.p.t. The basic plan of culture practice for the brackishwater ponds is more or less similar to that of open brackishwaters or nearshore area. However, these ponds should be fertilised in order to enhance the growth of seaweeds. Ammonium sulphate is mainly applied as a fertiliser source at 1 t/ha for the entire culture season of 3 months. Plastic bags (punched with very minute pores) containing required quantities of fertilisers are hung from culture frames. Such an arrangement would allow the nutrient solution to leach out slowly for the absorption of the growing seaweeds. Alternatively, the nutrient solution may be periodically sprinkled by manual labourers or be sprayed mechanically. The seaweed cuttings at 3-5 t/ha area are required and yields as much as 80 t/ha could be obtained from this unit area. Since seaweed culture unlike fish culture involves easy and inexpensive management phases and is lucrative, the unutilised brackishwater areas could be readily brought under this culture.

15

Bloodworms—Their Culture and Prospects

The larvae of *Chironomus* (Order: Diptera; Family: Chironomidae) are popularly known as bloodworms (Fig. 33). (The worm is 18-25 mm long and is characterised by the presence of four long ventral gills and two pairs of anal gills) which are blood red in colour. Unlike the adults (insects) which cause annoyance and allergies to some people, their larvae are totally harmless and they do not require blood meal. As bloodworms serve as a protein-rich live food source for fish fry in aquaculture systems and aquarium fish, the culture of these worms is a lucrative venture. While large scale culture of blood worms has been in vogue in many South-east Asian countries, it is rather new to India, although the adults are found abundantly in agriculture fields.

Culture Technique

Agriculture fields measuring 500-1000m^2 with 10-15cm depth could be converted into a culture area with bunds on all sides. Before initiating the culture operation, the pond is prepared by removing the grasses followed by sun-drying until the soil cracks. By this way, the eggs of parasites and predatory animals are eradicated. The ponds are then fertilised with dry chicken manure which is spread evenly on to the bottom @ 10 tonnes/ha. Alternatively, piggery manure, sheep manure, agriculture waste or super phosphate may also be applied. After a week of fertilisation, water from the nearby channel or well is pumped into the pond to a depth of about 10cm. This level has to be maintained throughout the culture operation. During this period, chironomid adults begin to lay eggs in masses which stick to the grasses at the edge of the field. After a week or 10 days, the larvae (0.3 cm in size) emerge from the eggs and nest in mud. To enhance the growth of these larvae, a mild insecticide is spread over the field @ 80 kg/ha after a week i.e. on 17th day of emerging larvae from egg masses. By 25th day, the larvae reach about 2 cm in length and are harvested. The harvest is done in morning or evening by a 1mm mesh net (diameter of net opening, 50cm; length, 150cm). The net is dragged by inserting the net opening perpendicularly into the mud to a depth of about 2-4 cm. With a view to clearing off debris, if any, the larvae collected are transferred to another net of mesh size, 1.5mm (diameter of net opening, 15cm; length, 90cm) which is immersed in a basin of water. Through the meshes of this net, the larvae wriggle out which are finally collected by yet another net of fine mesh (0.5mm) having a net opening diameter of 25 cm and length, 70 cm. The larvae are then frozen for marketing.

Courtesy : Fishing Chimes

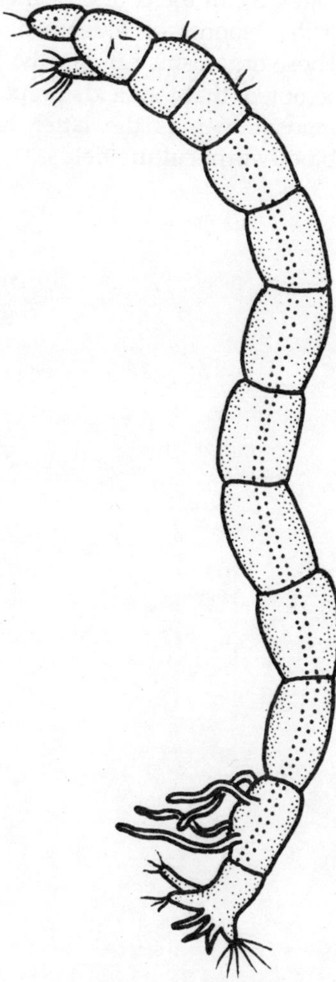

33

After the first harvest is completed, chicken manure is spread on to the field @ 2/tonnes/ha in order to compensate the nutrient loss. The second harvest is possible after about two weeks of the first crop. Depending on the growth and yield of larvae, a third crop of worms may also be harvested by adding the supplementary manure again. After the third harvest, the water of the pond is pumped out and the soil is sun-dried to facilitate the next cycle of operation.

Yield

In one cycle of about 50 days, an yield, of about 2 tonnes/ha is possible through three crops. As the cost of production of these larvae is very low and their price is high (Rs. 25/kg), a farming family of two adults can manage culture area of 5000-1000m^2 and get an annual income of about Rs. 50,000. The culture of bloodworms has added advantages, as, in the course of this culture, particularly when the column is more fertile, blooms of bluegreen algae or swarms of daphnias are likely to occur. These organisms can also be harvested for getting additional revenue. It should be noted that the fields prepared for the culture of bloodworms do not harbour mosquitoes as the latter hardly withstand the eutrophic water column of the bloodworm culture fields.

16

Water Recirculation Unit for Profitable Fish Culture

Space and water are two basic requirements, limiting the traditional fish culture practices in many countries of the world, as the demand for these is also increasing. In order to alleviate these problems and to bring areas of acute water scarcity or limited water supply under aquaculture, attempts are being made for re-using the water in closed and open re-circulation systems, which occupy less space as well.

The principle of this system is based on using a limited quantity of water for fish culture, by means of re-circulation and at the same time removing the metabolic wastes such as ammonia, carbondi-oxide, etc. An outdoor unit (see Fig. 34) consists of chiefly two reservoirs, an upper reservoir and a lower reservoir, and a number of fish culture tanks. Water from the upper reservoir is first exposed to ultra-violet light. The ultra-violet light destroys all the pathogenic microbes. The water then flows through a cascader (corrugated surface), to the different fish culture tanks by gravitation. During cascading, the water as it splashes is well oxygenated through atmospheric contact.

In fish culture tanks, desired water level is maintained with inlet and outlet arrangements through controlled water flow. The overflowing water from all these tanks is then collected in the lower reservoir where coarse particles are allowed to settle. The water from the latter is then pumped into the upper reservoir, one side of which is provided with a basal porous partition and a filter which is made of small pebbles.

These pebbles collect microflora such as *Nitrosomonas* and *Nitrobacter* and serve as an effective biological filter. These microbes oxidise the toxic ammonia derived from fish and convert it into non-toxic nitrites and nitrates.

In order to prevent overflowing as well as to safeguard a constant supply of water to the different reservoirs, the motor has to be controlled with a constant water level maintainer.

Advantages

The tanks of the unit could be conveniently converted into profitable fish nurseries where the stocking rate of fish fry could be tremendously increased and it could be as high as 25 million/ha.

The corresponding stocking rate in conventional cement cisterns is only 2 million/ha. Further, the survival rate of fish fry/fingerlings in this system could be 90% against 15 to 25% in conventional systems.

Courtesy : The Hindu

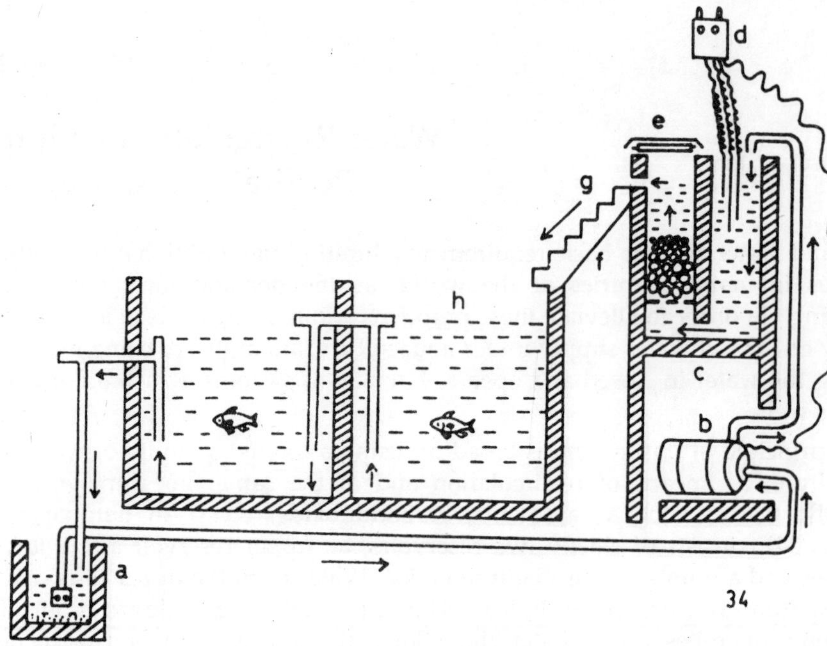

Collection or harvest of the fry or fingerlings, reared in this system is also easy and complete. If artificial feeding is to be sought throughout, water re-circulation system could also be a profitable unit for the culture of fish and prawns at high density and fish production can be enhanced by maintaining a stocking density of 0.5-1 million fingerlings/ha - a 50-100 fold increase than in conventional systems.

There is also much scope for the application of water recirculation system in brackishwater and marine areas for seed production and rearing.

17

Hydroponics and Fish Culture

Soilless agriculture known as 'Hydroponics' is the art of growing plants without soil in a solution of minerals and water. It is a valuable technique especially in areas where there is not enough fertile soil but the climate is conducive. In countries like USA, disease-free and edible crops such as tomatoes, potatoes, beets, carrots and greens and even roses are grown by this technique.

Integration of fish culture in hydroponic system is a recent one. Under this practice, fish and horticultural crops can be cultured in either aquarium tanks or in cement cisterns which are fertilised for the promotion of fish food organisms.

A typical hydroponic system (Fig. 35) for plant-fish culture consists of two compartments with a partition not touching the bottom at the centre. In one of the compartments, fish are grown and in other, a porous plate is shelved at the top touching the water column. Tender twigs of plants like *Bacilla rubra* are set on the pores of the plate. A control parallel system is also set up but without the fish. In this experiment, fish are fed with protein-rich (about 40%) pelleted feeds. Owing to the feeding of fish, the water of the tank is likely to get polluted due to the excretory products of fish such as ammonia, urea, uric acids and uneaten fish feed rich in nitrogen compounds. Among the nitrogen compounds, the unionized toxic gas viz. ammonia is formed directly through the excretion of gills and indirectly from various excretory products such as urea, uric acids and amino acids which are mineralised by the action of bacteria to form ammonia. Ammonia concentration beyond 0.5 mg/1 is said to be toxic to fish, as when the fish are exposed to high levels of ammonia, the haemoglobin of the blood loses the ability to combine with oxygen or to liberate carbon dioxide. It is reported that fish excrete ammonia and carbon dioxide at a concentration of 0-3-4.0 g/kg fish/day and 2-5 g/kg fish/day respectively. Further, the ammonia level is mainly influenced by pH rather than temperature and it is known to increase by 10 times at an increase of 1 unit in pH (say from 7 to 8).

In hydroponics fish culture systems, the growing plants utilise these waste products formed due to the metabolic activities of fish; thereby, not only fish and plants are harvested but also the tanks are self cleaned and stabilised.

Studies conducted with *Bacilla* in hydroponics-fish culture tanks showed that the growth of plants was more in experimental tank with fish than that of control tank without fish. This paves the way for expanding this technique in large water bodies where fish can be grown in the water column and horticultural crops be grown at the top of the water column. It is worth mentioning here that in the latter, the expenditure normally incurred towards fertilising the plants is

Courtesy : Kisan World

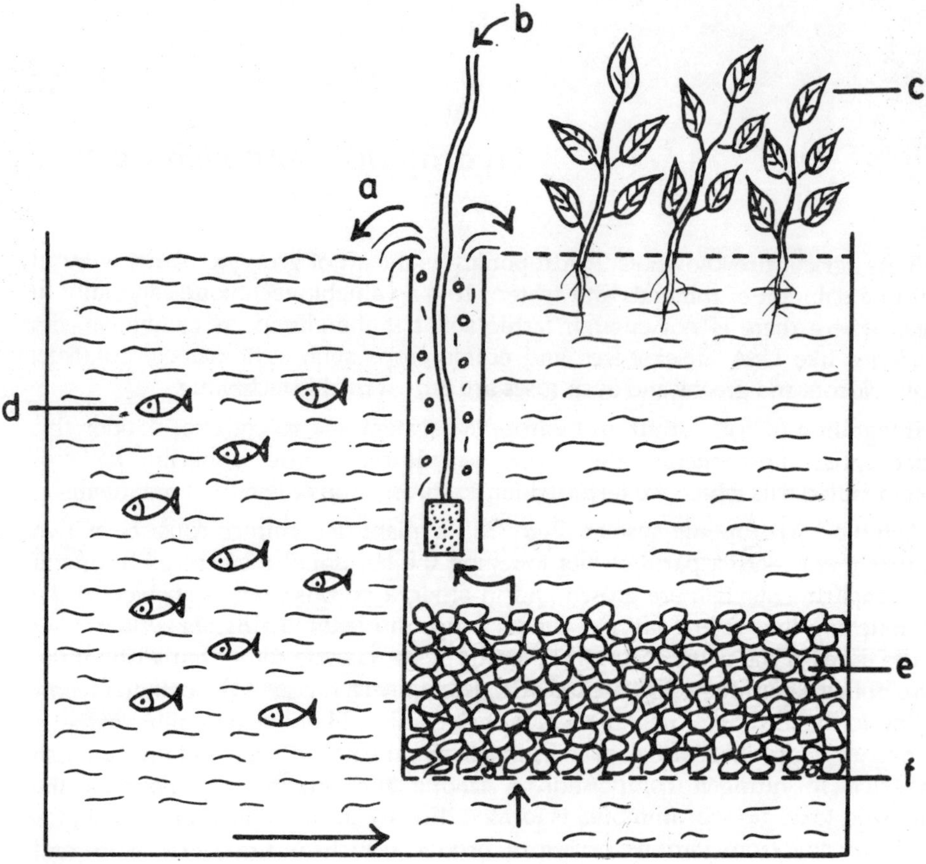

35

considerably curtailed. Further, this unique system of integrated soilless horticulture-cum-fish culture may be undertaken in small aquarium tanks in every house. The fish and plants apart from being of aesthetic value, could be periodically harvested to augment the supply of protein and other essentially required body building substances.

18

Non-Conventional Feeds for Profitable Aquaculture

In aquaculture, supplementary feeds constitute 50 per cent of the cost of fish production. The cost of available feeds is high (Rs. 3/kg) and generally, these feeds do not meet the actual protein requirements of growing fish or prawns. Hence, it is imperative to make use of the protein-rich and locally available non-conventional feeds.

Materials Identified: A number of non-conventional materials suitable in the preparation of fish feeds have been identified. The blue-green algae, *Spirulina platensis*, grown in sewage waters contain 40-70 per cent protein (on dry weight basis) and sufficient quantities of essential aminoacids such as lysine and tryptophen, vitamin B12, unsaturated fatty acids, carbohydrate and minerals. Unlike the cellulose cell wall of green algae, the mucoproteic constituents of the cell wall of *Spirulina platensis* are easily digestible.

Tapioca leaves have 20-40 per cent protein and a good amount of minerals and vitamin A. The toxic constituent Linamarin likely to be present in these leaves, may however, be removed by drying and boiling them.

Air-dried leaves of Subabul (*Leucaena leucocephala*), a recent addition to India contain 33 per cent crude protein and a variety of aminoacids similar to those in prawn waste and fish meal. The toxic mimosine content of the leaves is removed by heating the leaves at 80°C for two days.

Aquatic fern *Azolla pinnata* fixes almospheric nitrogen at the rate of 2-3 kg/ha/day owing to its symbiotic blue-green algae viz. *Anabaena azollae*. The dried azolla which has a crude protein content of 27 per cent also finds application in the feeds of pigs and poultry.

Mangrove leaves contain 8-18 per cent of protein in the decomposing state. The associated bacteria of the leaves are also known to increase the protein content besides making them easily digestible. Further, the bacterial flora may also improve the quality of food by providing essential aminoacids lacking as such in healthy leaves.

Seaweeds such as *Ulva fasciata*, *Enteromorpha intestinalis* and *E. compressa* (green); *Gracilaria corticata* and *G. foliifera* (red) and *Sargassum ilicifolium* (brown) have 15-25 per cent protein and a number of minerals to be included in fish feeds.

Other vegetable components are leguminous seed kernels, groundnut oilcake, rice bran, wheat bran, tapioca flour, and wheat flour. Non-conventional animal components include silk worm pupae, trashfish meal, prawn waste, squilla meal,

---

Courtesy : The Hindu

squid meal, chank meat, clam meat, pila meat and slaughter house waste. These have high protein content (50-70 per cent) and the inclusion of any one or two of these components is essential to enhance the protein content of feeds.

Ideal Proportion: The quality of feed determines the growth of fish and prawns. If the feed lacks optimal protein and is largely indigestible, then the assimilation efficiency and the resulting growth of fish would be poor. Hence, a complete and ideal feed with protein, fat, carbohydrate, vitamins and minerals at optimal levels is essential for making aquaculture practice lucrative.

For optimal growth, juvenile and adult fish and prawns need 30-40 per cent and 40 per cent protein respectively. A prawn feed containing 35 per cent protein may be prepared using animal component (50 per cent), groundnut oilcake (30 per cent) and tapioca flour (20 per cent) and a fish feed of 40 per cent protein with rice bran (15 per cent), groundnut oilcake (15 per cent), animal component (60 per cent) and tapioca flour (10 per cent). Cheaper feeds of varying protein levels could also be formulated and prepared with nonconventional components making use of their protein contents.

Preparation: The dried and powdered feed components are mixed and the mixture kneaded well adding about 30-50 per cent of water to form a soft dough. The dough is cooked for 30 minutes in steam under pressure at 1 kg/cm^2. The cooked dough is then fed through a hand pelletiser with a perforated disc having 1-2 mm holes or bigger holes depending on the size requirements of the feeds.

A semi-automatic pelletiser, powered by a 0.25 HP electric motor, suitable for the production of pelleted fish feed having a rated output of 10 kg/hr has been designed and developed by Tamil Nadu. Agricultural University, Coimbatore. The noodles are air-or sun-dried and broken into pieces of about 1cm. Pelleted feeds with more than 15 per cent moisture lose stability in water and are prone to the attack of fungus.

19

HCG Induced Breeding in Freshwater Fishes — A Boon to Fish Farmers

The success of fish culture depends mainly on the year-round supply of seed fish. In nature, the maturation of gametes and associated spawning in fish are largely influenced by variations in environmental factors such as light, water temperature, water level of the habitat, pH, salt concentration, etc. The synchronisation of egg production and ovulation by broodfish have been achieved to a greater extent by inducing them with fish pituitary extracts containing the hormone, 'Gonadotropin'. Though the technique of 'hypophysation' is in vogue in India since 1957, a substitute for fish pituitary glands was sought owing to the latter's poor availability and exorbitant cost (Rs. 250/g).

Breeding in freshwater carps and catfish have been successfully done by inducing them with a unique stimulating hormone viz. 'Human Chorionic Gonadotropin (HCG) present in the urine of pregnant woman. Unlike fish pituitaries, highly purified and potent preparation of HCG with a standard biological activity of 30 IU/mg is readily available as a commercial product in 500 mg polythene packs at a cheaper rate (Rs. 70/g). The HCG can be administered alone or in conjunction with fish pituitary extracts in breeder fish.

Preparation of HCG and Pituitary Extracts: The HCG solution is prepared by macerating the required quantity of HCG powder in a homogeniser containing known volume of water for about five minutes. The solution is then centrifuged and the supernatant light pinkfish fluid is taken for use. The extract of the freshly collected and alcohol-preserved pituitary glands is also prepared in the same way. The extract is invariably prepared at a concentration of 10 mg/ml. Both the solutions are mixed together and taken in hypodermic syringe for intramuscular or intraperitoneal injection in the breeder fish. (Fig. 36)

Application of HCG and Pituitary Extract: About two days before induing the fish to breed, healthy, disease-free and fully ripe male and female breeders are selected. The females are first given a preparatory dose of HCG at the rate of 4-5 mg/kg body weight. This dose advances the development of gonad in females. After two days of administration of maturing dose, the breeding dose is decided depending on the quality of spawners especially their age, size and gonadal development. Before injection, the breeders are anaesthetised using either quinaldine (1:1000) or MS222 (1:20,000). The females are given two injections at the rate of 8mg/kg and 4mg/kg respectively maintaining an interval of three hours. The males on the other hand are given only a single dose at the rate of 5 mg/kg fish at the time of second dose to females. The maximum amount of fluid that is injected at one spot, however, should not exceed 0.5 - 1 ml/kg body weight. The hormonal extracts for the use of the above may be 70%

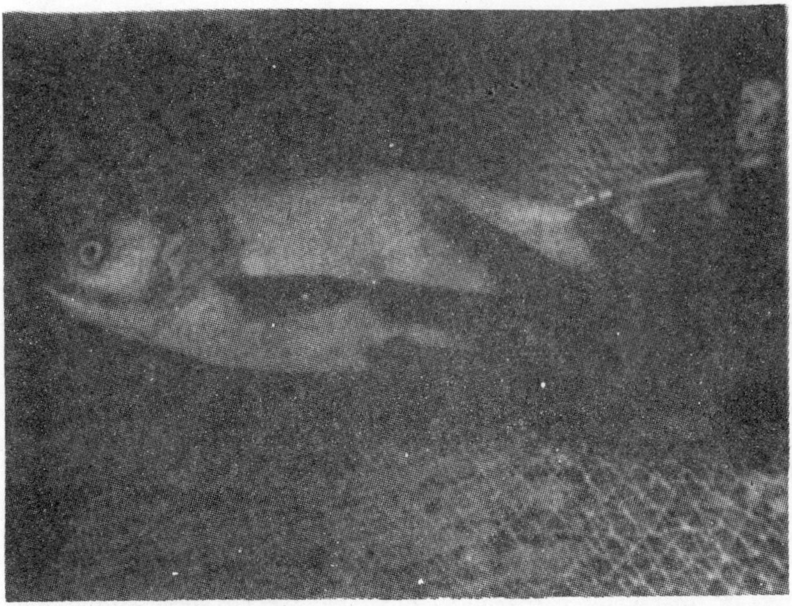

36

37

HCG and 30% fish pituitaries. The breeders after injection are introduced in the breeding hapa, (Fig. 37) a rectangular cloth chamber installed in the water column. The spawning occurs about 3 hrs of final injection and the fertilized eggs are then collected and transferred to the hatching hapas. (Fig. 38)

Application of HCG Alone: Periodical application of HCG in Indian major carps such as catla, rohu, and mrigal and catfish such as singhi and maghur is found to help both in the early maturation of their gonads and growth. Immature fishes 1-2 months before breeding may be injected with HCG at 5-10 mg/kg body weight one or three times depending on age and condition of fish. This would make them mature quickly. Further, before the administration of breeding drugs, if fairly mature fishes are injected daily with HCG (6 mg/kg body weight for female and 3 mg/kg body weight for male) for about a week, this would hasten their milting or ovulating process. In silvercarp, in which self absorption of eggs without ovulation is a common phenomenon, breeding has been reported after 15 days of HCG injection. Two doses for females at 7 mg/kg body weight and 10 mg/kg body weight with an interval of 6 hrs and one dose for male at 5 mg/kg body weight have yielded better results in silvercarp. Further HCG -treated fish can be used more than once for induced breeding in the same season.

Since the technology of HCG - induced breeding is cheaper and easier to handle, this could be a commercial proposition of seed production in freshwater carps. However, though the use of HCG is common among the fish farmers of West Bengal, it is yet to gain momentum in other States.

20

Guppy—A Unique Ornamental Fish

Guppies belong to the species *Poecilia reticulata (=Lebistus reticulatus)* of the Family Cyprinodontidae, and have been named after an aquarist, Robert John Lechmere Guppy who had sent a few living pairs of this species to the British Museum, London in 1860. As guppies are collected in large numbers from freshwater bodies, they also bear the name 'Millions Fish'. Though the guppies had their origin in Trinidad, West Indies, they are grown in many countries as ornamental and larvicidal fish to combat diseases due to mosquitoes and their larvae. However, knowledge of the biology of this species especially its reproduction and genetics is scanty.

Sexing of Guppy

The females of guppy measure 6 cm and males 3 cm. The former are plain-olive or greyish brown in colour and the latter decorated with countless colours. According to their colouration, the males are named red types, green types, tuxedos, callico patterns, mosaics, fire works, king cobra, albinoes, etc. The females are also characterised by wide variation in their caudal fin and the presence of a tube shaped copulatory organ, gonopodium, which is the modified anal fin. Based on variations in tail, the important varieties of guppy are roundtail, topswordtail, bottomswordtail, lyretail, pintail, veiltail, doubleswordtail, deltatail, lacetail, scissorstail, bettatail, spadetail, or cofertail, flagtail, speartail, tuxedotail, fantail, kingcobratail, and Robson roundtail (Fig. 39 A-R). Round or squaretailed strains are however most common. Gregor Mendel's laws of heredity are not only effective but also clearly observed in the guppy. Unfortunately, almost all the attractive characters are recessives in that they are obscured in the first generation. The dominant and recessive characters in guppy are recorded in Table 1.

Table 1: **Dominant and Recessive Characters of Guppy**

Dominant characters	*Recessive characters*
Wild grey	Gold
Wild grey	Blond
Wild grey	Albino
Round tail	Pointed tail
Round tail	Veiltail
Round tail	Swordtail
Normal caudal fin colour	Black blotches on caudal fin
Normal or stunted growth	Gigantism

Courtesy : Science Reporter

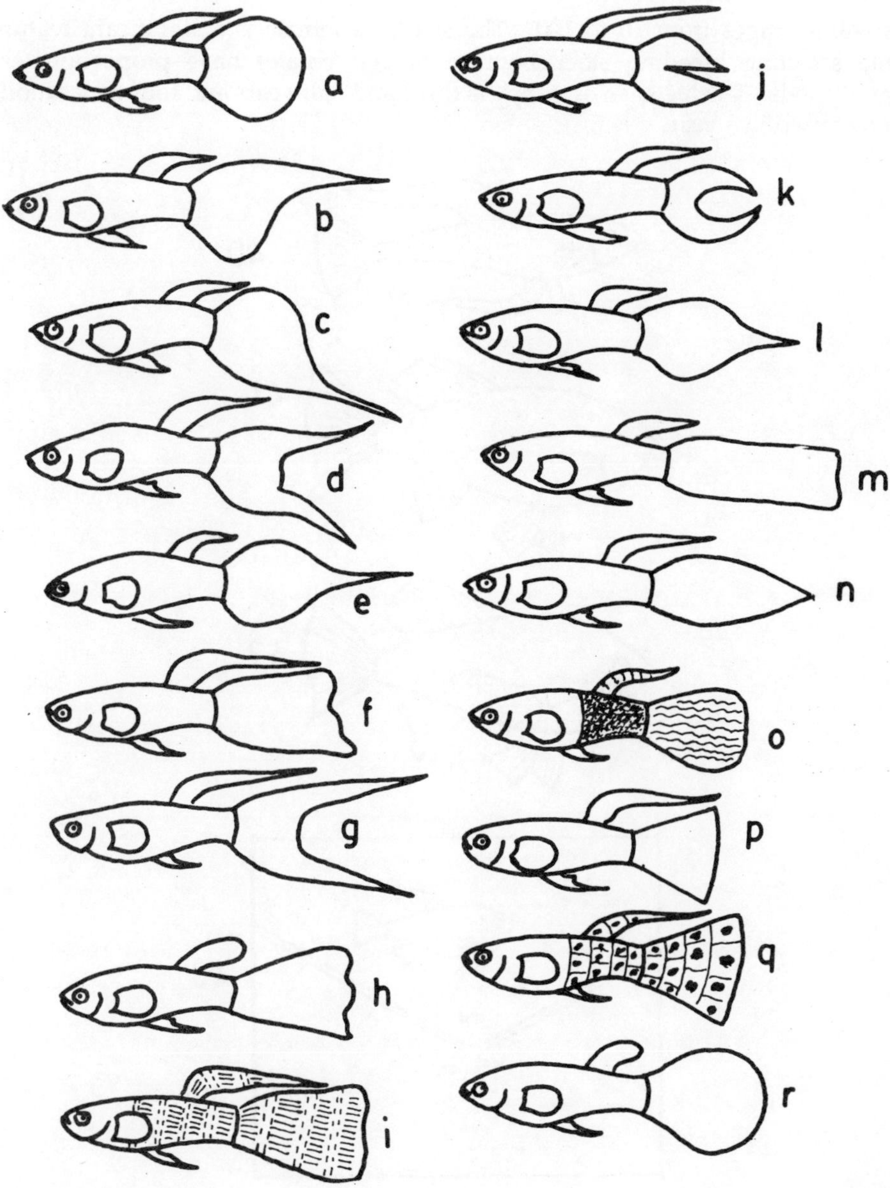

39

Reproduction

The female attains sexual maturity in three months. Fertilization is internal. The male (Fig. 40) impregnates the female (Fig. 41, 42) which carries eggs in an incubating pouch until they hatch as tiny fish. The gestation period is 3 to 4 weeks depending on temperature. The brood female is identified by a dark coloured gravid spot in front of anal fin which is fan shaped. When the colour of this spot deepens, birth of young ones is imminent. The number of fry per spawning ranges from 20 to 100. The size of a female is a significant feature while selecting breeding stock because bigger females have proportionately larger broods. The life span of guppy is three and half years and four generations can be raised in a year.

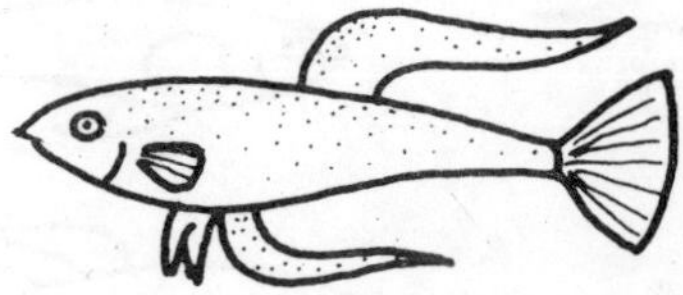

40

41

42

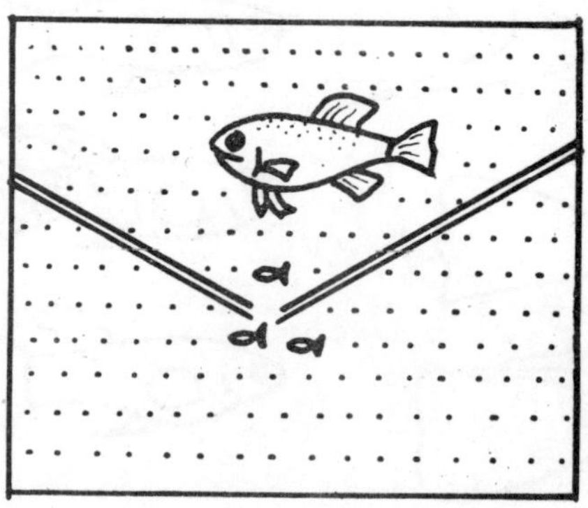

43

Breeding

Guppies are bred successfully in a community tank in which different species of ornamental fish live. However, the tank must be densely planted if the fry are to be protected from the parent which is known to devour its own young immediately after their release. Female brood guppy can also be transferred to a breeding trap (Fig. 43) which is provided with a partition open at the base. In its original form, the male guppy is pastel in colour with few dark spots and line which, breeding through many generations, has led to certain mutations. Many new and beautiful types of guppy with varying types of tail and dorsal fin shapes have come into being. However, it is almost impossible to find two similar looking males even when the fish are brothers from a long line of closely inbred ancestors. It is desirable to inbreed brothers and sisters or backcross father and daughter or mother and son to get desired results. Though other fishes such as swordtails, platys and Siamese fighters have been crossed amongst themselves for a wide range of attractively coloured hybrids, crossing of these fish with guppy is not possible owing to difference in their chromosomal numbers. For example, guppy possesses 46 chromosomes against 42 of Siamese fighters.

Conditions of Aquarium Tank

Compared with other species of ornamental fish, guppies are very hardy. Under conditions which no other species could sustain, the guppies flourish and breed incessantly. As they are peaceful fish and mix with others, they are ideal for community tanks. The water of the tank should be moderately hard and slightly alkaline. The best temperature range for maintaining guppies healthy is 25°C-27°C. The adult guppies are fed with a variety of prepared and fresh live foods as well as vegetation. The fry are fed chiefly with the nauplii of brine shrimp.

Exposing female guppies to tobacco smoke is known to cause sterility. In the case of pregnant guppies, deformed fry are produced. Development of spinal cord may also be affected when their diet is without vitamin C. Vitamin A deficiency in diet leads to blindness, stunting and haemorrhage affecting the base of their fins. Excessive intake of fat, on the other hand, leads to lipoid liver degeneration and anaemic fins. Guppies are also infested with flukes in which the first sign is dropsy and subsequent development of small swellings on the body. Owing to swim bladder deformities, adult guppies may also develop curved bodies.

21

Biological Filters and Air-lift Pumps for Aquaria

Maintaining the quality of water in aquarium tanks is indispensable for growing ornamental fish. Excreta of fish and unused food materials often lower the oxygen content of water in tanks. Unless water is changed at least once in two days in aquarium tanks, optimal survival of fish cannot be expected. Home aquarists are therefore forced to make alternative arrangements for changing water in tanks when they are not at home for a longer time. Biological filters and air-lift pumps are nowadays used in aquaria to help aerate and maintain the quality of water. These modern devices are simple in design, low in cost, easy to maintain and efficient.

Biological filter (Fig. 44) consists of suitably cut and perforated asbestos sheet (or fibre glass or acrylic plate) which is fixed at about 10 cm above the bottom of aquarium tank using rubber corks as supports at its four corners. This sheet fits tightly in the tank and serves as a false-bottom. Above the plate, washed gravels are spread uniformly for a thickness of about 5 cm.

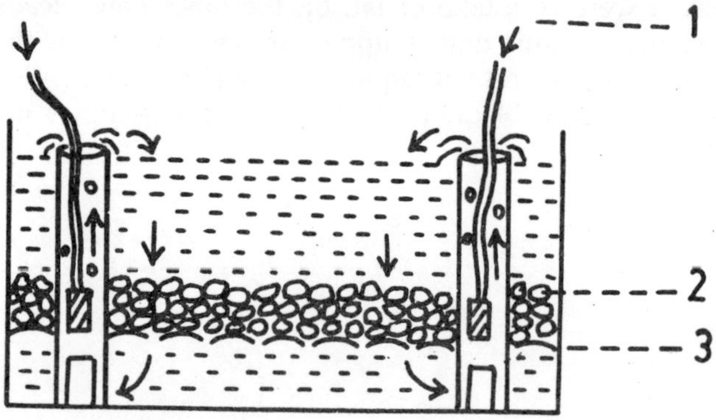

44

Water in an aquarium is liable to contamination especially with ammonia due to the metabolic activities of fish. When the ammonia-laden water passes through the gravel bed, it is activated by aerobic nitrifying bacteria such as *Nitrosomonas* and *Nitrobacter* present in it. The *Nitrosomonas* oxidises ammonia into nitrite,which is further converted into nitrate by *Nitrobacter*. The ammonia-free water so formed occupies the bottom of the tank, i.e. below the false-bottom and is subsequently air-lifted to the surface.

Courtesy : Science Reporter

Air-lift (Fig. 45) is a vertical pipe made of PVC or fibre glass. Its diameter and length vary with the size of an aquarium tank. The top and bottom of the pipe are open. The bottom edge of the tube has three rectangular slits cut out at equidistant points on it. In an aquarium tank, the pipe is held in holes of the false-bottom in such a way that the mouth of the tube is at about 1 cm above the water level. An air-stone connected to an aerator or compressor through a polythene tube is kept inside the pipe.

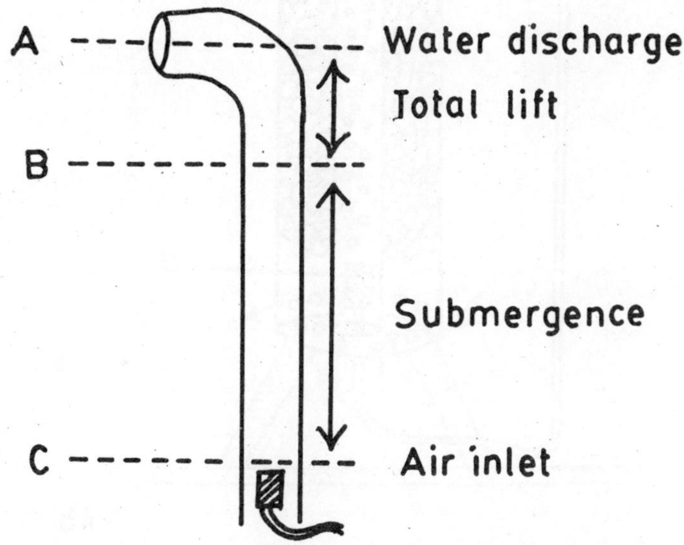

45

It is known that when a pipe is held vertically in water, the level of water inside and outside the pipe is equal. When air is injected at the lower end of the pipe, it rises up and forms bubbles. It produces a mixture of air and water which is lighter than water alone. The air-water mixture inside the pipe is therefore lighter than the water outside the pipe and so the prevailing equality of water levels is disturbed. Heavier water from underneath the filter plate moves into the lower end of the pipe. As long as air is injected into water, the air-water mixture is spitted out at the top of the pipe. The length of the lift pipe held below affects the efficiency of an air-lift. An air-lift works less effectively when the volume of the injected air exceeds the capacity of the lift pipe.

An air-lift pump with in-built biological filter unit (Fig. 46) is composed of two dissimilar cylindrical plastic tubes, which open at both ends. The smaller tube (diameter, 2.5 cm and length, 30 cm) is held firmly inside the larger one (diameter 7.5 cm and length, 32 cm) with the help of three acrylic plastic struts. The upper opening of the inner tube is about 2 cm lower than that of the outer. The entire unit stands on three plastic legs on the floor of the aquarium. The space between

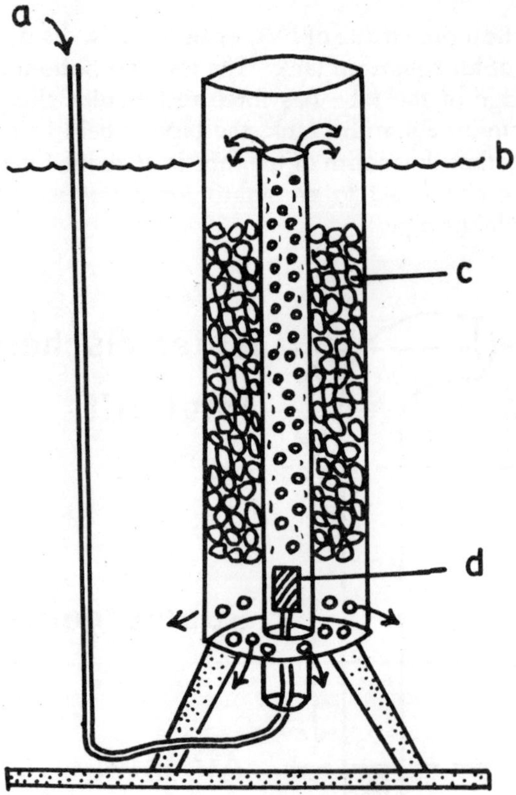

46

the inner and outer tube is filled with gravels and the water level in the aquarium is maintained at a level slightly lower than the upper end of the inner tube. An air-stone is kept at the base of the inner tube. The upward displacement of air forces the water to spill over into the outer tube. The gravel bed of the latter serves as a biological filter.

22

Balanced Diets for Aquarium Fishes

The dietary requirements of aquarium fish are different from those of food fish. A diet for aquarium feeding should be nutritionally balanced, palatable, resistant to crumbling, water stable, buoyant and inexpensive. A well balanced diet besides yielding higher growth also provides the nutrients necessary to hasten recovery from diseases or help the fish in overcoming the effects of environmental stress.

The diet of fish should normally have protein, carbohydrate, fat, essential aminoacids, vitamins and minerals in right proportions so that, it meets the daily energy and nutrient requirements to support the maintenance and growth of fish. While the carbohydrate is the chief energy yielding substance, much of the protein is converted into flesh during the growth of fish. The following table shows the utilization of feed by an aquarium fish during its metabolism.

Intake (*g*)					*Loss* (*g*)	*Growth (flesh formation)* (*g*)
Protein	:	30-40	Faeces	:	20-40	
Carbohydrate	:	30-70	Branchial & urinary loss	:	10	
Fat	:	10-20	Heat (maintenance, activity, etc.)	:	25-45	
Total		100			75	25

Aquarium fish depending on their feeding habits and preference may be fed with livefoods such as mosquito larvae, fruit flies, bloodworm, cyclops, tubifex, daphnia, rotifers, brine shrimp and earthworm or with moist pellets, dry pellets, flakes, and chopped fish, shrimp, beef, oyster, crab, liver, spleen, lung and heart. The live or wet foods are expensive and lead to poor water quality and pollution. Dry or flake feed on the other hand is easy to manufacture, transport and use. Most nutrients in dry feeds including vitamins and unsaturated fats are stable at room temperature and therefore such feeds could be stored safely without freezing. Further, feeding dry pellets or flakes either by hand or using mechanical feeders is also much simpler than that of the wet feeds.

Feed Formulae

The dietary protein requirements are size dependent i.e. smaller fish require higher levels of protein for maximal growth than larger fish. Generally, aquarium fish need feeds containing 30-40% protein. However, carnivorous fish feed should have higher levels of fish meal and fish oil in order to improve its acceptability and increased intake. An ideal 40% protein feed suitable for aquarium fish should have the following ingredients.

Major ingredients

	For herbivorous fish (%)	*For carnivorous fish(%)*
A.		
Fish meal	25	30
Shrimp meal (Wastes from head and shell)	22	10
Soybean meal	25	20
Rice or wheat bran	20	25
Tapioca flour (binder)	5	10
Fish oil	3	5
	100	100
B. Essential Aminoacids (g/kg dry feed)		
Arginine	25	
Histidine	5	
Isoleucine	10	
Leucine	15	
Lysine	20	
Methionine & Cystine	15	
Phenylelenine	20	
Threonine	10	
Tryptophan	2	
Valine	15	
C. Vitamins		
(i) *Fat soluble vitamins*		
Vitamin A (IU/kg feed)	3500	
Vitamin D_3 (IU/kg feed)	3000	
Vitamin E (IU/kg feed)	100	
Vitamin K (mg/kg feed)	10	
(ii) *Water soluble vitamins (mg/kg feed)*		
Ascorbic acid	400	
Vitamin B_{12}	0.02	
Biotin	0.4	
Choline	3000	
Folic acid	5	
Inositol	100	
Niacin	150	
Pantothenic acid	60	
Pyridoxine	20	
Riboflavin	20	
Thiamine	10	

Contd.

D. Minerals

Calcium (g/kg feed)	0.2 - 0.3
Inorganic phosphorus (g/kg feed)	5-10
Magnesium (g/kg feed)	0.5
Copper (mg/kg feed)	3
Manganese (mg/kg feed)	10-15
Selenium (mg/kg feed)	0.1 - 0.4
Zinc (mg/kg feed)	15 - 30
Iodine (ug/kg feed)	0.5 - 1
Iron	< 1
Aluminium	< 1
Chromium	< 1
Vanadium	< 1

Preparation of Dry Pelleted Feeds

The required quantities of finely powdered dry materials excepting fish oil, vitamin - mineral mix are put together and kneaded well adding minimum quantity of water to form a soft dough which is cooked for 30 minutes in steam under pressure at 15 1b/inch^2. After cooling, to the cooked dough, the fish oil and vitamin - mineral mix are added and the mixture is thoroughly stirred. The dough is then fed through an extruder having a perforated disc with 2mm dia. holes in a hand pelletiser (Fig. 47). The noodles are dried either in the sun or in an artificial drier and broken into pieces of required length, usually 5 to 10 mm. Feed pellets of various dimensions could however be prepared to suit the feed size requirements of fishes of different sizes.

Preparation of Flaked Fish Diets

All the dry ingredients without fish oil and vitamin - mineral mix are mixed and the mixture is ground in a mill so that 90% passes through a 60 mesh sieve. The fish oil is then added. Hot water (90°C) is added to the mixture to adjust the solid content to 30% and all the materials are mixed in a high speed blender for three minutes. Vitamin - mineral mix is then added to this and the mixture is further blended for 30 seconds. The purees are flaked by spreading them in a layer approximately 1mm thick on a flat metal surface maintained at 150°C by suspension in oil. The dried flakes which are approximately 0.2 mm thick are removed from the heated surface after 20 seconds, cooled for 5 minutes and packed in hermetically sealed flexible pouches within 30 minutes. Flake foods are preferred by 80% of world aquarium keepers as they disperse well in aquarium tanks.

23

How to Protect Aquarium Fishes from Diseases

Aquarium fishes, owing to high stocking densities and poor water quality in tanks are likely to be affected by many diseases which include infestations with protozoan and metazoan parasites, bacterial and fungal infections, neoplasius (unwanted growth to tissues) and nutritional abnormalities. If the diseased fishes before their release into the tank are not quarantined and properly treated, besides their death, other fishes in the tank will also be affected leading to huge loss and discouragement to even enthusiastic aquarists.

How to Identify Diseased Fish

The following symptoms may indicate an aquarist that his fishes in the tank are diseased; (i) refusal of food; (ii) closed fins (dorsal fin being the first to close); (iii) lean body; (iv) swelling of stomach; (v) erection of scales; (vi) appearance of fungal patches; (vii) arrest of movement to surface or bottom; (viii) loss of sense of balance and direction; and (ix) appearance of external bumps, sores or spots.

Common Diseases of Aquarium Fishes

1. *Fin* or *tail rot* (Fig.48). It is caused by certain species of bacteria which attack the fins or any injured areas of skin.
2. *Dropsy.* It is also a bacterial disease affecting mainly Siamese fighters, mollies and gouramis. The affected fish are swollen due to the accumulation of excess fluid in the body.
3. *Cotton-wool disease.* It is caused by the fungus *Saprolegnia* which appears mainly on raw wounds of Barbus, Danio, etc. The first sign is a few short white hairs protruding from infected spots. Subsequently, these lengthen and spread over a large area. Besides affecting the movements of fish, the secretion of mucus is also affected.
4. *White spot disease* (Fig. 49). It is recognised by the appearance of white spots of 1 mm size on the skin and gills. The causative organism is a minute protozoan, *Ichthyophthirius* which burrows under the scales of fish and forms cysts when favourable conditions occur. These cysts burst open in 3-4 days releasing hundreds of young parasites.
5. *Costiasis* (Fig. 50 & 51). It is caused by a small protozoan flagellate, *Costia necatrix,* which attacks the skin and gills of fish. A reddish or bluish streak along the side of the fin from the tail towards the head is the primary symptom of this disease.

Courtesy : Science Reporter

48

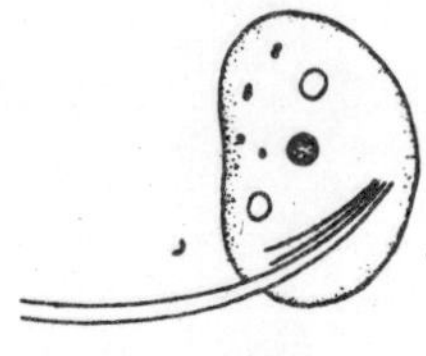

50

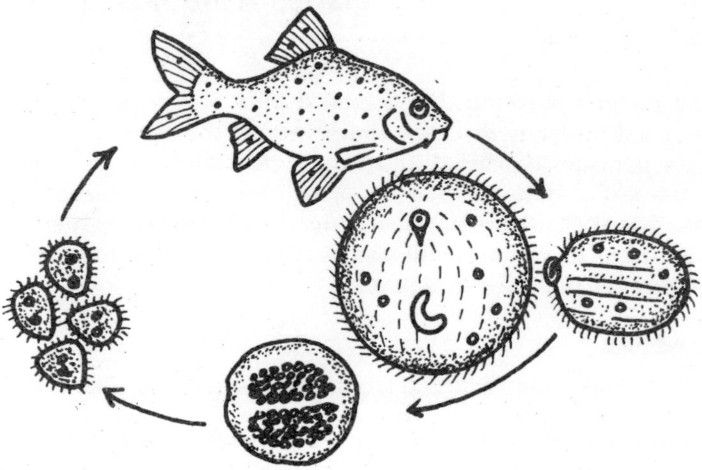

49

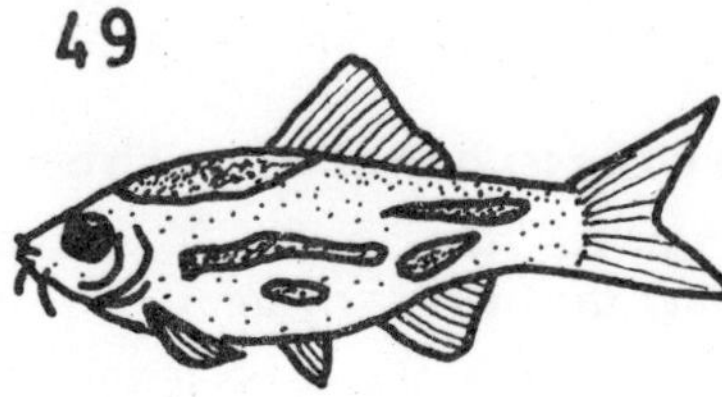

51

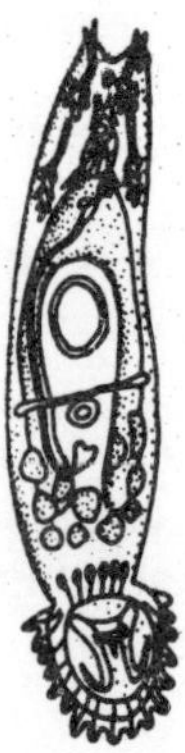

53

52

54

55

6. *Velvet disease.* It is due to the protozoan *Oodinium limneticum* and is characterised by an yellow-brown film on the back of the fish. The edges of fins and gills are largely affected.
7. *Trematode flukes.* Flatworms *Gyrodactylus* (Fig. 52) and *Dactylogyrus* (Fig. 53) often lodge in the gills and skin of guppies. The affected fish swims in a wild and jerky manner.
8. *Anchorworms.* Free-swimming crustaceans such as *Lernaea* (Fig. 54), *Lernaenicus* (Fig. 55, 56) and *Caligus* (Fig. 57,58) attack mainly large fishes such as cichlids and kissing and giant gouramis. As these lodge under the skin, their egg sacs are only protruding.
9. *Fish louse* (*Argulus* sp-Fig. 59). Likewise anchorworms, it also attacks larger aquarium fishes. The presence of spine between eyes is an important feature as it injects poison especially in the belly, gills and throat.

Table 1: **Treatments for Various Diseases in Aquarium Fishes**

Parasite/disease	*Treatment*
1. Fin rot	1 minute bath in 500 ppm copper sulphate
2. Dropsy	Bath in 0.7% salt water (however, there is no permanent cure)
3. Cottonwool disease and white spot disease	(i) 24 hr bath in 0.15 ppm malachite green+25-50 ppm formalin (ii) 3% hydrogen peroxide or 0.1% potassium dichromate
4. Costiasis	1 hr bath in 1.5% salt water or 200 ppm formalin
5. Velvet disease	30 minute bath in 1% methylene blue
6. Trematode flukes and external protozoans	24 hr bath in 25 ppm formalin or 10% ammonia solution
7. Anchorworms, fish lice and leeches	30 minute bath in 300 ppm formalin; painting the affected area with 0.1% potassium permanganate or turpentine followed by removal with a pair of tweezers.
8. Gas bubble disease	Transfer the affected fish to normal water

10. *Leech* (Fig. 60). Certain species of leeches attack the body of larger aquarium fishes and live on their blood.
11. *Gas bubble disease.* Strong plant growth and the resulting supersaturation of dissolved oxygen often lead to the appearance of fine gas bubbles under the skin. As the liquid in the body of the fish is also getting supersaturated, the movements of the fish are largely affected by this disease.

In aquarium fishes, infections of internal organs are probably of lesser significance than external diseases.

56

57

58

59

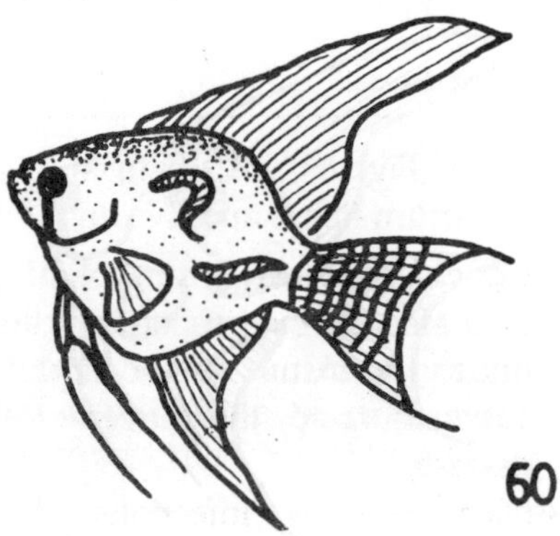

60

Controlling Measures

Quarantine. In order to prevent the diseases and the external parasites, all fishes brought from the shop are first given a 30 minute bath in a quarantine tank containing 1-2 ppm potassium permanganate solution. Alternatively, the fishes may be isolated in a quarantine tank containing about five litres of water, one teaspoonful of salt and five drops of 5% aqueous solution of methylene blue.

Treatment

Although single treatment is sufficient for ectoparasites, repeated treatments are indispensable for eliminating cysts embedded in the host tissues. The simplest treatment for the affected freshwater aquarium fishes is to transfer them to salt water as many parasites have narrow salinity range. Parasites with a complex life cycle are however excluded by removing their intermediate stages in the parasitic life cycle. The treatments/prophylactic measures for various diseases are given in Table 1.

24

Mass Culture of Zooplankton for Coastal Aquaculture

The right type of livefoods such as zooplankton and *Artemia* are essential for feeding the larvae of fish and crustaceans in the hatcheries. As these organisms possess rich protein, high reproductive rate, short regeneration time and the ability to live and grow in crowded conditions, their mass culture would be a profitable venture. Among the different species of zooplankton, the brackishwater rotifers *Brachionus plicatilis* and cladocerans such as *Moina* spp. and *Daphnia* spp. (Fig. 61,62) have received much attention in recent years.

61

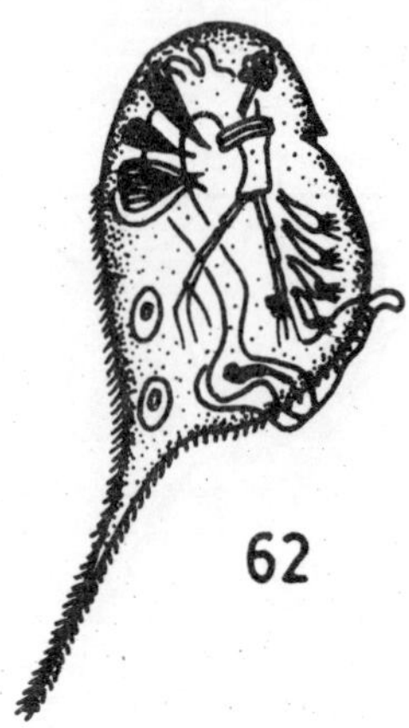

62

Culture of *B. plicatilis*

This species is considered to be an excellent food item of fish and prawn larvae. The culture of *B. plicatilis* is invariably undertaken in glass carboys or tanks (100-20000l) made of fibre glass or concrete. The tanks are first filled to two-thirds with filtered seawater diluted with chlorine-free tap water or well water, the salinity of which is maintained at 15-35%. It is also necessary to aerate the above culture tanks with compressed air in order to keep the oxygen level at more than 6 mg/1. As *B. plicatilis* multiplies rapidly by parthenogenesis, the egg-bearing females of *B. plicatilis* are normally isolated from a brackishwater pond and are kept in a series of petridishes. Suitable live feeds are given to these organisms until their eggs hatch into young ones which are used for inoculating the prepared culture vessels at a density of 1000 organisms/1. The growing rotifer larvae are fed with microscopic algae such as *Chlorella, Tetraselmis, Dunaliella* and *Spirulina,* the cultures of which are maintained separately. The growing rotifers may also be fed with yeast and algal powder at 200 mg/larva/ day. Continuous aeration and 15 hours illumination per day are also necessary to ensure good growth of rotifers under controlled conditions. *B. plicatilis* attains a density of 500-1000 per ml of culture medium within a week time. It is advisable to harvest the rotifers when they carry eggs, since maximum organic content of the rotifers is noticed at this stage. It is also better to harvest one thirds of the growing rotifers and the volume harvested is replaced with fresh algal culture. Such periodical harvesting would keep the rotifer culture in good condition for a longer period. The harvested rotifers are concentrated by passing through a nylon bolting silk cloth of 40-90 um mesh size which removes insect larvae, nematodes or detritus, if any. The concentrated rotifers are then washed in filtered seawater for use either in fresh or frozen condition. The frozen rotifers at -40°C can be held up to one year for use as food for prawn larvae.

Culture of Cladocerans

Freshwater cladocerans like *Daphnia* and *Moina* serve as ideal live food sources for fish fry and post larvae of prawns. Adults and parthenogenetic female cladocerans collected from freshwater impoundments serve as a stock culture. The culture vessels (100-2000l) made of fibre glass, plastic or concrete are first filled with chlorine free tap water and are fertilised with urea, superphosphate and groundnut oilcake at the rate of 5kg, 3kg, and 200 kg per m^3 of water respectively. After two days of fertilisation, the water in these tanks becomes greenish and are ready for inoculation with stock culture of *Moina* or *Daphnia* at 1-2 number/ml. The growing cladocerans are fed with the green algal species of *Chlorella* and the culture tanks are aerated and illuminated as for rotifer tanks. The cladocerans multiply rapidly and attains a density of about 20,000/1/week.

The harvesting of cladocerans is done when the females reproduce actively by parthenogenesis. Parthenogenetic females are considered to be more nutritive than females without eggs or males. As in the case of rotifer culture, one third volume of water alongwith the growing cladocerans is harvested first and the volume is compensated by adding suitable volume of freshwater alongwith fertiliser mix. The concentrated cladocerans are also processed similar to rotifers and used.

25

New Techniques in the Transport of Fish and Prawn Seed for Aquaculture

Transport of fish and prawn seed from the areas of their collection or procurement to culture site is an important phase of aquaculture. If it is not properly planned and implemented, large scale mortality of seed would occur resulting in heavy loss. New techniques replacing the old traditional ones such as lifting in mud pots and hundies help largely in this venture.

Anaesthetisation of fish seed

As the fish seed are very active, they should be anaesthetised before their transport so that, the available dissolved oxygen of the medium will not be exhausted at a shorter time. Nowadays several anaesthetics find application in the transport of fish seed. Though the concentration of various anaesthetics used differs widely depending on the species, size, etc., they have a common characteristic feature, viz. lowering the metabolic activity of the fishes by their depressing action on the brain leading to the low consumption of oxygen in the media during transit and facilitating higher survival and long duration of journey in oxygen-packed containers under high stocking densities. Any anaesthetic that is used for the transport of fish seed must possess the following; it must be easily soluble; the dosage required should be low; the time of induction and recovery should be short; the treated fish should tolerate well for several hours at low concentrations; it should not have any side effects in the fish and the lethal concentrations should be high, so that, the fish do not die accidentally.

The commonly used anaesthetics include MS 222 (Tricaine methane sulphonate) (0.3 ppm); quinaldine (2 methyl quinoline) (0.1 ml/1) and carbonic acid (500 ppm). As quinaldine is not soluble in water, the solution is made using acetone. For the preparation of 500 ppm carbonic acid solution, 8 ml of 7% sodium bicarbonate is added to 8 ml of 4% sulphuric acid.

During anaesthetisation, the fish seed are given a bath of any of the above anaesthetics until they lose their equilibrium and remain on the bottom. As soon as the seed are removed from the anaesthetic solution, they are washed before transferring to the transportation containers. Generally, fingerlings weighing about 1 kg are transported for 6 hours in 2 litres of water containing MS 222 which is added directly to the medium at 1:100000.

Transport of Fish Seed in Oxygen-packed Containers

The anaesthetised fish seed can also be transported in oxygen-packed containers which facilitates stocking of a large number of seed per unit volume of water at a known duration of journey. Aluminium or galvanised iron tins of 40 x 25 cm size and polythene bags of 65 x 45 cm size are often preferred for the transport of fish seed (Fig. 63). To each tin, a pore-proof polythene bag is kept inside and about one-thirds of its capacity is filled with filtered and aerated water, the salinity of which depends on the area of fish seed collected. Then the calculated number of one-day starved and acclimated fish seed are carefully released inside the bag. A tube from the oxygen cylinder is inserted into the bag and the upper part of the bag about 6-8 cm is twisted, folded and a string is kept ready for tying. The oxygen is then slowly drawn at 5-8 1bs/sq. inch from a nearby cylinder through the tube until the two-thirds of the bag is inflated or the height of the inflated bag equals the height of the tin. Simultaneously, the air from the innerside of the bag is expelled. As soon as the inflated condition of the bag is obtained, the string is secured and the tin is closed (Fig. 64), and thus the tins for transport are made readv. In order to enhance the oxygen level of the medium, a few ml of oxidising agents such as potassium permanganate or hydrogen peroxide solution may also be added.

Transport of Prawn Seed

Limited quantity of prawn seed for short distances could be transported without aeration. For example, 50 juvenile prawns or 250 postlarvae of less than 20 mm size may be transported in one litre of water up to two hours without aeration. On the otherhand, the transportation of 25 juveniles or 100 postlarvae in a litre of water may be facilitated by changing the water twice. However, for long distances which involve much transist time, oxygen packing of seed is indispensable in order to reduce their mortality. Before oxygen packing, the prawn seed are conditioned for 2-3 hrs as described already. It is also worth mentioning here that the seed of the white prawn *Penaeus indicus* are more delicate than that of the tiger prawn, *P. monodon*. The number of seed which can be transported under oxygen packing with reference to duration of transport is given below:

Size of postlarvae/juveniles (mm)	*Duration of transport* (hrs)	*No./1*
8 - 12	24	300
- do -	36	150
14 - 18	24	250
- do -	> 24	100

For the surface transport of prawn seed for more than 12 hrs, the oxygen-packed containers are kept inside the card board boxes or tins. For 20 hr air transport of *P. monodon* seed, about 2000 individuals may be oxygen packed in a tin of about 20 litres. Aquatic weeds or inert artificial plastic weeds may also be put inside the containers for providing shelter to the moving prawn seed. The juvenile prawns should always be transported in thick polythene bags or in suitable containers as their sharp rostrum may damage flimsy bags.

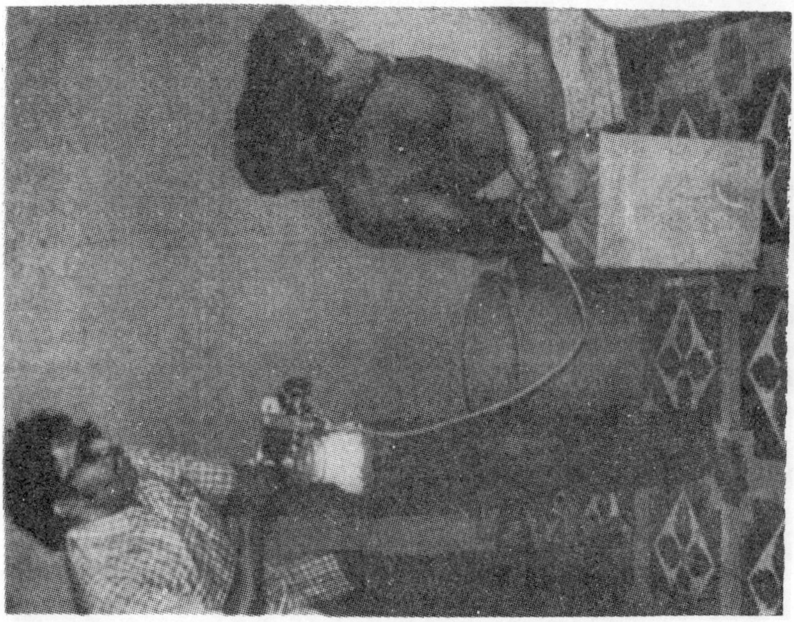

Determination of Quantity of Fish and Prawn Seed

The quantity of seed for transport depends on their size, mode and duration of transport, salinity of the medium and ambient temperature. A simple formula to determine the optimal quantity of.seed is given below:

$$N = \frac{(DO - 2) \times V}{c \times h} \text{ where,}$$

DO, dissolved oxygen in ambient water (mg/1); V, Vol. of water (1); c, rate of oxygen consumption by individual fish (mg/hr); h, duration of transportation (hr).

General Guide Lines

(i) Before transport, fish and prawn seed are conditioned by starving them for about a day in a hapa installed in running water systems. This would facilitate in the removal of excreta of fish seed which would otherwise pollute the medium and lower the oxygen level during transit.

(ii) Before packing and transport, the fish seed should be treated with 1% potassium permanganate solution or common salt solution which serves as a prophylactic measure.

(iii) Transport of seed must be done during cool hours especially during night or early morning.

(iv) During journey, transporting containers should be wrapped with wet gunny bags in order to reduce the temperature. Alternatively, ice blocks and saw dust may be kept inside the container which reduces the metabolic rate and oxygen consumption of seed thereby increasing their survival rate.

(v) Jolting of containers during transit should be avoided as this would injure the seed and make them die.

(vi) Before stocking in culture ponds, the transported seed should gradually be acclimated with pond water.

26

New Methods of Preservation of Fish Gametes for Aquaculture

The fish stocks of the world aquatic systems show a decreasing trend owing to increased exploitation and therefore aquaculture is considered to be the only hope of meeting the protein demand and mitigating malnutrition. The periods of maturity and spawning of culturable fishes and that of water availability (monsoon seasons) do not synchronise and hence the seed requirements of fish farmers are not completely satisfied. In order to overcome such constraints and boost the fish and their seed production through aquaculture, the storage of fish gametes is essential as it facilitates artificial reproduction and preserves desirable gene pools.

The fish gametes may be stored either by short term storage of unfrozen gametes or long term storage of frozen gametes by 'cryopreservation'. While the former ranges from hours to weeks and is generally undertaken in hatcheries to overcome asynchromy in maturation, transport of gametes or selective use of spawners, the latter helps in the indefinite storage of desirable genes to increase the efficiency of selective breeding of culturable species as like that of domesticated animals. Further, it is beneficial particularly when female and male gametes are collected and used at different times or locations or when the collection site and incubation facility are some distance apart and delayed fertilization is necessary.

Compared to the eggs, the sperm cells are preferred for preservation owing to their large number, ease of collection and suitability for cryopreservation. Further, the fish spermatozoa are immotile either in the testis or in the seminal plasma. During natural reproduction, their motility is induced after the discharge of sperm into the aqueous environment or in the female genital tract. As there is no expenditure of energy in fish spermatozoa before fertilization their life span is considerably increased besides facilitating successful storage from a few hours to several days.

Extenders (diluents) and cryoprotectants: These agents are helpful in the preservation of fish gametes in many ways. The diluents such as seawater (17-18 ppt), sodium chloride (8.8%), urea (0.3%) and sodium bicarbonate provide better control of the physiochemical conditions during storage besides extending the storage time and maintaining higher level of fertility. The cryoprotectants such as dimethyl sulphoxide (3-10% v/v) and ethylene glycol reduce the nucleation temperature of both the suspension medium and cell liquid and allow the storage of cells in a supercooled state at several degree below 0°C. Further they act by altering the chemical and physical properties of the extracellular medium rather than by affecting the cells directly. Therefore, they protect against injury from slow freezing.

Short term storage of fish spermatozes: In this process, the spermatozea of fish are diluted with any of the above mentioned extenders and are stored from 0.8° to 4°C for up to three weeks.

Supercooling of fish spermatozoa: In this process, the spermatozoa are stored in temperatures between -3° and -6.5°C in the presence of any cryoprotectant.

Post-mortem storage: The spermatozoa of herring species are known to show high fertility for over two days when the dead fish are kept at 0.8°C. Similarly, the spermatozoa of chum salmon show higher viability when the dead fishes are maintained at 11-12°C.

Cryopreservation of fish gametes: This method is found to be suitable for silvercarp, greymullet and Indian major carps in general and rohu, grasscarp and milkfish in particular. The spermatozoa of these fishes are first diluted with any of the extenders such as sodium chloride or sodium bicarbonate and then treated with a cryoprotectant. Afterwards, the freezing of spermatozoa is done with either dry ice (-79°C) or liquid nitrogen (-196°C) followed by thawing at the time of use.

Freezing and thawing: These are the most important parameters in the preservation of fish gametes. The optimum freezing rate of treated fish sperm in carps may be about 10°- 20°C/min and thawing rate from 120°C/min to 1500°C/min so as to have a post thaw fertility at 90%.

During the preservation of fish gametes, the following should also be taken into consideration.

(i) The mode of fertilization (external or internal) as well as the spawning environment (sea, brackishwater or freshwater) should be recorded before the storage of fish gametes.

(ii) Sperm collected from several males during the peak spawning season should be pooled and used as they respond best. It is advisable to chill the milt immediately and to process it as quickly as possible.

(iii) Usually the cryoprotectants viz. dimethyl sulphoxide and glycol are more suitable for marine and freshwater fish respectively as the spermatozoa of marine fish are generally heavier than that of freshwater fishes.

(iv) In order to determine the optimal freezing rate for the spermatozoa of different species, various ranges between 1 and 5°C/min, 10 and 50°C/min and above 100°C/min may be tested.

(v) As the duration of motility and fertility in cryopreserved cells can be extremely short, immediate insemination after thawing is necessary.

(vi) Generally for better viability, aged spermatozoa should not be used for preservation.

27

Artificial Seawater

Like many brands of instant tea and coffee, synthetic sea salts are also available in many countries. These salts are used for preparation of artificial seawater to grow fish and other invertebrates. Due to the installation of industries near seashore environments and due to their discharge of chemical effluents into the adjacent seas, natural seawater in many countries has become unfit for biological and other studies. Natural seawater due to its higher concentration of organic matter and flourishing growth of ciliates, fungi and bacteria is unfit for the culture of fish food organisms and also for the rearing of baby fishes. Further, the transport of natural seawater to distant places is a very costly affair. Considering all these, artificial seawater is a best source.

A good quality artificial seawater should not lack trace elements or contain major elements in abnormal ratios. Artificial seawater prepared in correct ratio of major and minor elements can alone support life. Though there are several formulas for the preparation of artificial seawater, one which is widely used is given in Table 1.

Table 1: **Composition of Artificial Seawater (Chlorinity, 19.38)**

Salts	*Weight/litre*
Major Salts	
Sodium chloride	24.5336g
Magnesium chloride	11.1120 g
Sodium sulphate	4.0944 g
Minor salts	
Potassium chloride	0.6945 g
Sodium bicarbonate	0.2011 g
Potassium bromide	0.1005 g
Boric acid	0.0272 g
Calcium chloride	0.0348 g
Sodium fluoride	0.0030 g
Trace elements	
Barium nitrate	0.0994 mg
Manganous nitrate	0.0546 mg
Cuprous nitrate	0.0396 mg
Zinc nitrate	0.0157 mg
Lead nitrate	0.0066 mg

Courtesy : Science Reporter

While preparing artificial seawater, all the major, minor and trace salts should not be mixed simultaneously as it may lead to precipitation and make the medium unfit for use. Hence, major salts should be mixed first in tap water and stored in a suitable container for a day with aeration. During the second day, minor salts are added and the third day, trace salts. On the fourth day, the pH of the solution should be adjusted to 8.5 by adding 0.1 molar solution of sodium bicarbonate. The solution is ready for use during the fifth day. Invariably the prepared artificial seawater is stored with a drop of formalin solution or 0.1 ppm mercuric chloride solution in order to minimize any microbiological interferences that might develop as a result of exposure of saline solution to air over long periods of time.

Nowadays mariculture operations are gaining much importance throughout the world. The success of the mariculture depends mainly on the availability of healthy seed of fish and shell fish and fish food organisms like brine shrimp, rotifer, worms, etc. Such organisms could be cultured in large quantities in artificial environments using synthetic seawater. Although many brands of instant sea salts are readily available in U.S.A. and other developed countries for the preparation of artificial seawater, in India we have not yet thought of this for immediate use because our nearshore waters are free from pollution to a greater extent. However, artificial seawater may find its application in the near future.

28

Sea as a Source of Drugs

Sea has long been known as an enormous store-house of minerals and food materials. Recent researches have discovered that organisms of sea could act as receptacles for the extraction of many new drugs some of which may be beneficial in treating cancer and others for use against mental illness or common cold. The different drugs and their sources are given below.

Antibiotics: Extracts of marine sponges particularly of Jamaica or Caribbean seas possess antimicrobial, anti-fungal, Ichthyotoxic and anti-tumour substances. The species *Verongia gigantea* has bromine containing antibiotic substances. Still some are known to have the drug Ara-c (arabinosyl cytosine) which could serve as a therapeutic agent against certain cancers. Similarly many corals belonging to the Phylum Coelenterata also possess anti-bacterial substances.

Elatol: Red seaweed contains a compound 'elatol' an anti-cancer drug as it inhibits completely the cell division.

AP-A: The gram sea anemone *Anthopleura xanthogrammica* contains a powerful cardiac stimulant viz. Anthopleurin-A (AP-A), an enormously complex polypeptide with a chain of about 90 amino acids. It is 300 to 500 times more powerful than 'digitalis', the commonly used heart stimulant in medicine. Furthermore, AP-A does not have the side effects of digitalis. AP-A could be extracted in tiny amounts (15-45 parts per million) and it is not present at all in very large and very small specimens.

Anticoagulant: There is no acceptable substitute now for 'heperin', the anticlotting agent in human blood extracted from the lungs of cattle and intestinal mucosa of swine, such anticoagulants are present in sharks and discarded wastes of crabs, lobsters, clams and scallops.

Chitosan: It is derived from the chitin, a chemical component of carapace of many crustaceans. It is found to have amazing wound healing properties and could be useful following surgery. Chitin derivatives have many other potential uses. In paper industry to increase its wet strength and to improve its printing surface and, in water purification processes, in agriculture with insecticides (since it resists biodegradation it would provide long-term time release effectiveness on the insecticides).

Coenzymes: The coenzymes in the liver of lemon shark have been found to accelerate the body's defence mechanisms. Experiments conducted with rats and mice have proved that when these animals were treated with shark liver extract, there was a significant increase of phagocyte production.

Courtesy : The Hindu

Toxins: Many dinoflagellates (single celled algae) are known to release neurotoxins during metabolism. These toxins work by interfering with sodium metabolism of humans. As drugs that cause reactions with sodium are rare nowadays, the above toxins if properly modified and controlled could be of considerable value in treating many human diseases.

29
Economic Importance of Diatoms

The fertility of the sea on which the growing human population depends for protein-rich fish and shellfish is determined mainly by the unicellular, microscopic plants called 'diatoms' (Fig. 65), a chief component of phytoplankton forming the basic link in the food chain of the sea. Though a majority of them lead a floating life on the surface of the sea, species like *Cocconeis ceticola* are epizoic and cling to the body of the whales. Owing to this, the colour of the whales turns to yellow and this strange phenomenon is known as 'sulphur bottom'.

The size of the diatoms is microscopic, varying from 0.005 to 1.2 mm and each cell has two valves resembling a pill box. The ornamentation of the siliceous valves with radially or bilaterally arranged openings or striations is interesting and is of taxonomic value. The solitary, colonial or chain-like species possess projections, long setae, wing or keel like protrusions for increase in surface area for floatation.

The total annual production of diatoms of all seas has been estimated as 1.2 to 1.5×10^{10} tonnes. Besides serving as a chief food to many of the herbivorous fish and shellfish species like *Fragilaria oceanica* and *Hemidiscus hardmannianus* are known to indicate the abundance of foodfish like sardines and choodai in the Indian west coast and east coast respectively. Similarly, *Fragilaria antarctica* indicates the abundance of the krill, the chief crustacean food source of baleen whales, seals and squids in the polar seas. The upwelling regions or the most productive west coasts of all seas, wherefrom, about 75% of the fish are caught, are especially rich in diatom populations.

Diatomin, the food reserve of diatoms is rich in vitamin A and seems to be the precursor for the vitamin A content of fish oils. This hydrogen compound is also said to be the precursor for the natural petroleum deposits of the deep seas. When the diatom populations of an area die in toto, their frustules (shells) form a bed called 'diatomite' 'diatomaceous earth' or 'Kieselghur'. The latter, being an indicator for the presence of petroleum, is also exploited for use in commerce. The diatomite with pure silica is used primarily as an aid in the filtration of fruit juice, syrup and varnish; in the boilers, electric ovens and refrigerators as insulators; as polishing agent for precious metals; in the separation of paraffin wax from petroleum and in the manufacture of concrete where 1 to 3 kg of diatomite is mixed with 50 kg of cement.

Though the importance of diatoms and diatomite is well known in developed countries, developing countries are yet to undertake large scale investigations in their seas in order to locate these resources for profitable exploitation.

Courtesy : Fishing Chimes

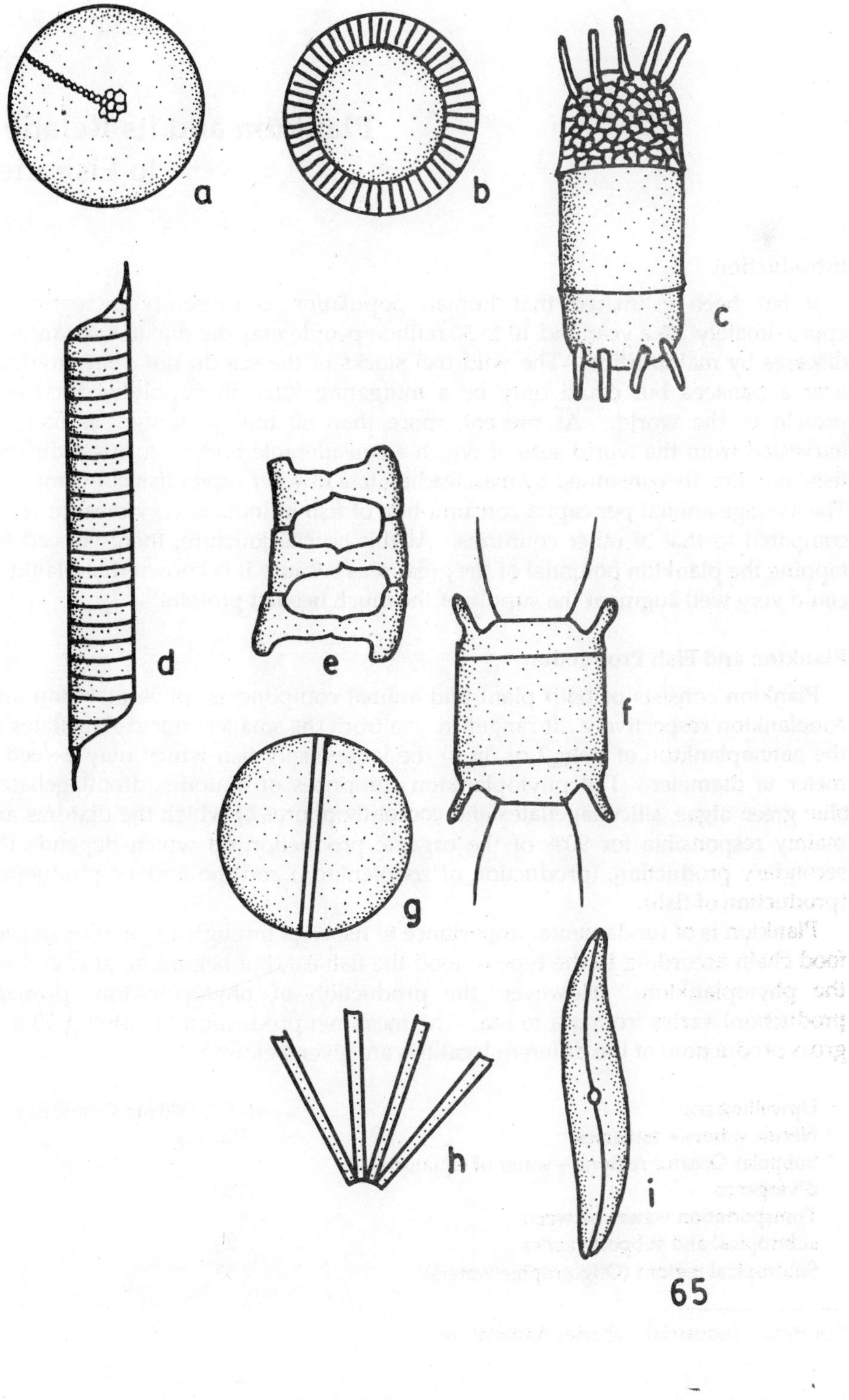

65

30

Plankton and its Relation to Fisheries

Introduction

It has been estimated that human population is presently increasing at approximately 2% a year and 10 to 30 million people may die due to starvation or diseases by malnutrition. The wild fish stocks of the sea do not offer anything near a panacea but could only be a mitigating force in supplying increased protein to the world. At present, more than 60 million tonnes of fish are harvested from the world seas of which a considerable portion forms industrial fish, not directly consumed by man leading to a low per capita fish consumption. The average annual per capita consumption of fish in India is very meagre when compared to that of other countries. At this crucial juncture, there is need for tapping the plankton potential of the great seas because it is known that plankton could very well augment the supply of the much needed protein.

Plankton and Fish Production

Plankton consists of both plant and animal components, phytoplankton and zooplankton respectively. It ranges in size from the smallest micro-flagellates in the nannoplankton of only 2 or 3*u* to the largest jelly fish which may exceed a meter in diameter. The phytoplankton comprises of diatoms, dinoflagellates, blue green algae, silicoflagellates and coccolithophores of which the diatoms are mainly responsible for 90% of the organic production on which depends the secondary production, (production of zooplankton) and the tertiary production (production of fish).

Plankton is of fundamental importance to fisheries through a longer or shorter food chain according to the type of food the fish eat, but beginning always with the phytoplankton. However, the production of phytoplankton (primary production) varies from sea to sea. The mean net production (assuming 50% of gross production) of the different localities are given below:

Upwelling sea	1000-10000 mg	$C/m^2/day$
Neritic waters + estuaries	500 mg	"
Subpolar Oceanic regions + water of equatorial divergence	100	"
Transportation waters between subtropical and subpolar zones	70	"
Subtropical regions (Oligotrophic waters)	35	"

Courtesy : Industrial Fisheries Association

In a few restricted areas of the world, particularly along the west coasts of the continents of subtropical latitudes where there are prevailing off shore winds and strong eastern boundary currents, surface waters are diverted off shore and are replaced by nutrient-rich deeper waters. Such areas of coastal upwelling are biologically the richest parts of the ocean. They exist off Peru, California, northwest and southwest Africa, Somalia, Malabar Coast, Orissa coast and in other more localized regions. Extensive coastal upwelling is also known to occur in various places around Antarctica although its exact location and extent have not been well documented. During periods of active upwelling, the primary production exceeds 1.0 and may reach even $10g/Cm^2/day$. The upwelling areas of the world in the order of primary production are ranked below.

$g C/m^2/day$	Upwelling areas
>1	Peru, Canary, Benguela, S.Arabia, Vietnam, Gulf of Thailand, etc.
0.3-10	Somali, Malagasy, Orissa coast, Java, Sri Lanka, etc.
0.3	California, Chile, New Guinea, Andaman Islands, N.W. Australia, etc.
0.1	Open Oceans

The tertiary production in fish and other nektonic production is derived by taking 1% of the primary production of 10% of the secondary production. Based on this, the fish production of some of the important upwelling areas is depicted here.

Peru and Benguela	12-18	million tons wet wt/yr
Canary	8	"
California	3	"
Costa Rica dome	2.6	"
Off Java 2.3 Southern Arabic	2.5	"
Somali	1.25	"
Malabar coast of India.	1.2	"
Orissa coast of India	1.0	"
Ghana	1.0	"

From the point of view of the commercial fisheries, the upwelling areas are much more important than the open ocean because the intensity of production in $g C/m^2/day$ even with reduced transfer coefficient will yield three times as much material to the herbivores. Further, pelagic herbivorous fishes like herrings and anchovies subsist on the colonial, large size phytoplankton species such as *Chaetoceros* spp., *Skeletonema costatum*, *Asterionella japonica*, *Schroederella delicatula* and *Fragilaria* spp., directly in the upwelling area, whereas in the open ocean, the primary herbivores or phytoplankton-eating microzooplankton are fed first by macrocarnivorous zooplankton or smaller fishes which are finally eaten by top carnivores like tuna, dolphin and squid. Thus the number of trophic levels in the upwelling area is only less than 2 compared to that of the oceanic or coastal provinces where the corresponding values are 5 and 3.

Next to the upwelling areas, the coastal waters are the most productive. The levels of primary production and the estimated fish production of the oceanic, coastal and upwelling areas of the world are given in Table 1 & 2.

Table 1: **Division of the Ocean with Provinces According to their Levels of Primary Organic Production (After Ryther, 1971)**

Province	*% of Ocean*	*Area (Km²)*	*Mean Productivity (g C/m³/yr)*	*Total Production (10⁹ tons of C/yr)*
Open ocean	90	326×10^6	50	16.3
Coastal zone	9.9	36×10^6	100	13.6
Upwelling areas	0.1	3.6×10^5	300	0.1
			Total	20.0

Table 2: **Estimated Fish Production in the Three Provinces from Primary Production (After Ryther, 1971)**

Province	*Primary Production (tons of organic C)*	*Trophic level*	*Efficiency*	*Fish production (tons-fish wt.)*
Oceanic	16.3×10^9	9	10	16×10^5
Coastal	3.6×10^9	3	15	12×10^7
Upwelling	0.1×10^9	$1\frac{1}{2}$	20	12×10^7
			Total	24 x 107

It is worth mentioning here that the upwelling areas which form only 0.1% of the total oceanic area contribute nearly half of the world fish production. The other half is derived from the coastal waters and few off shore regions of comparatively high fertility. Though 240 million tons of fish (fresh weight) are produced annually in the seas of the world, we are able to harvest only about 60 million tons of fish. Hence, it is imperative to explore the new fishing areas and unexploited commercially important species so as to enhance the harvestable yield.

Plankton as Biological Indicators

The importance of plankton as animal food is well known. The whale shark, *Rhincodon typus*, the biggest fish measuring 20 m in length feeds exclusively on zooplankton. Apart from food value, several species of plankton are known to indicate the water masses thereby, influence the distribution and the abundance of fish and shellfish. Among diatoms, *Fragilaria oceanica* and *Hemidiscus hardmannianus* are known to indicate the abundance of oil sardine in the west coast and choodai in the east coast of India respectively. Similarly, the 'white water' phenomenon due to the abundance of coccolithophores is also known to be a good omen for herring fishery.

Most of the dinoflagellates emit light (cold living light) which lasts for about 1/10 second and is equal to mechanical, chemical or other stimulations. Fishermen often locate schools of sardines and other shoaling fishes by the luminescence of the dinoflagellates.

Among zooplankton, the arrow worms, particularly *Sagitta elegans* and *S. setosa* are known to indicate the presence of mixed and unmixed waters in the North sea and English channel respectively. The appearance of *S. elegans* in the North sea is always found associated with the inflow of Atlantic waters resulting in good fishery. On the other hand, when *S. setosa* is present in the channel water, young fish are scarce. The association of copepods particularly *Calanaus* spp. with rich herring shoals is also worth mentioning. Similarly, the colonial diatom, *Fragilariopsis antarctica* indicates the abundance of the antarctic krill, *Euphausia superba* which in turn helps the fishermen to locate the abundance of baleen whales, seals and squids. This relationship is a direct one as *F. antarctica* and krill form food for krill and baleen whales respectively.

Plankton as a Product of Commerce

Majority of the diatoms possess diatomin, a food reserve in the form of oil which aids in floatation. The diatomin is rich in vitamin A and hence the vitamin A content of fish oils is due to the feeding of fishes on daitoms. The diatom oil, a hydrocarbon compound is said to be the chief source for the formation of natural petroleum. When the diatoms die, they sink to the bottom and form a bed called 'Diatomaceous earth'. If such a bed is not disturbed by currents and other related water movements, this develops into areas of rich petroleum grounds. Scientists suggest that for the petroleum tanks of Venezuela at Los Angeles, the diatoms were largely responsible.

Diatomaceous earth is also nowadays used in industries. It is employed in the filtration of fruit juice, syrup and varnish; in the boilers, electric ovens and refrigerators as insulators; as polishing agent for precious metals; in the separation of paraffin wax from petroleum and in the manufacture of concrete, where 1 to 3 kg of diatomaceous earth is mixed with 50 kg of cement. Of late diatomaceous earth is incorporated in paints for illumination especially during darkness for road and other safety signals.

Plankton as Human Food

Apart from its commercial use, plankton could be a valuable human food source. However, it is the cost of extraction of plankton that makes it at present a doubtful commercial proposition. For example, 5 million gallons of water have to be filtered to get the plankton equivalent to 0.5 kg of beans of our kitchen gardens. However, in arctic and antarctic areas where the euphausid krills are dominant, the fishing of krills is a profitable venture. It is suggested that krills may one day rival soybeans as a source of protein. The annual production of krill has been estimated as 200 million tons which is more than 2 times the world fish catch. However, the annual krill catch at present is only 20,000 tons. Hence there is a vast scope for expanding the fishing operations particularly in the vicinity of krill-rich grounds. Krill could be a good source for the large scale preparation of 'Fish Protein Concentrate, (FPC), an easily digestible direct human food. Further, it is also incorporated in the poultry and other livestock feeds.

The high protein content of plankton (60% of the dry weight) besides lipids (15%), carbohydrates (2-3%) and ash (20%) make it a nutritious food item. Dried copepods could be conveniently converted into human food and is better than beef and other animal flesh. It is also interesting to record here that the liquid of the cell-content contains less salt than seawater.

Certain spices of mysids are collected for food in East Indies. Similarly, a deep water copepod namely, *Euchaeta norvegica* is a delicacy which one day might support a fishery for luxury markets. Although collection of plankton as human food is not going to be a profitable venture excepting arctic and antarctic krills, plankton from other latitudes, even though of less magnitude is a good potential source for pharmaceutical industries and investigations on this line are indispensable.

Plankton as Hindrance

The formation of 'red tide' or discolouration of water due to the sudden outburst of certain species of dinoflagellates especially *Gymnodinium breve, Gonyaulax polyedra, G.catenella G.acatenella, G.tamarensis* and *G. monilata* has caused serious concern to fisheries. The mass mortalities of fishes due to either neurotoxin excreted by dinoflagellates or depletion of oxygen following the death of dinoflagellate cells have been recorded from the coasts of India, Southwest Africa, Japan, Southern California and Gulf of Mexico. The red tide causing dinoflagellates may also be responsible to paralytic shell fish (especially oysters) poisoning as the shell fish are known to filter these dinoflagellates in high densities. It is worth mentioning here that a single serving of infected shell fish containing only a few milligram of toxin is lethal to man.

Certain zooplanktonic predators like jelly fishes, arrow worm and the copepod, *Euchaeta* spp. may voraciously feed on fish larvae resulting sparse fish population. It is also interesting to record that waters of rich jelly fish are invariably without fish and further, the operation of fishing nets in such waters is also almost impossible due to the clogging of the nets. Investigations have also revealed that the catch of herrings showed an inverse relation with arrow worms and the pteropod, *Limacina* spp.

Certain phytoplanktonic organisms particularly, *Rhizosolenia* spp. when abundant turns the water into a soup called 'Dutchman's baccy juice', by which not only herrings are sparse but also the operation of fishing nets is problematic.

Conclusion

Eventhough a good amount of work on plankton is available in India and elsewhere, its direct or indirect influence on the individual fishery is not known. Hence, an indepth study on plankton is very imperative. Furthermore, ways and means to detect the rich planktonic grounds like Deep Scattering Layer have also to be formulated to tap the planktonic resources. As stated earlier, the pharmacological values of plankton also need the immediate attention of our research workers. A thorough knowledge on this line would go a long way in improving the quantity of many life saving drugs which could pave way for earning the foreign exchange.

31

Tintinnids (Marine Protozoans) as Fishery Indicators

Though information on the marine protozoans is plentiful, detailed studies on an important and hitherto neglected group of planktonic ciliates, viz. 'tintinnids' (Phylum Protozoa; Class: Ciliata; Order: Tintinnida) were almost nil in India till 1970.

Out of the 800 species of tintinnids so far known, more than 98% are confined to marine environment. They are cosmopolitan in distribution and are restricted to the surface layers as planktonic organisms. The size of tintinnid species may vary from 50 um to 500 um in length and from 20 um to 150 um in diameter. It is worth mentioning here that the tintinnids dwell in houses or 'loricae' secreted by them. These houses may be tubular or pot like or vase shaped (Fig. 66-83). The lorica may be agglomerate or clean. When it is agglomerated, it is strengthened by foreign particles like small stones, diatom frustules and tests of foraminiferans and radiolarians. The ciliated body of the tintinnids attaches to the bottom of the lorical cavity through stalk. However, in species where both the ends of the lorica are open, the attachment is on any one of the inner sides of the lorica.

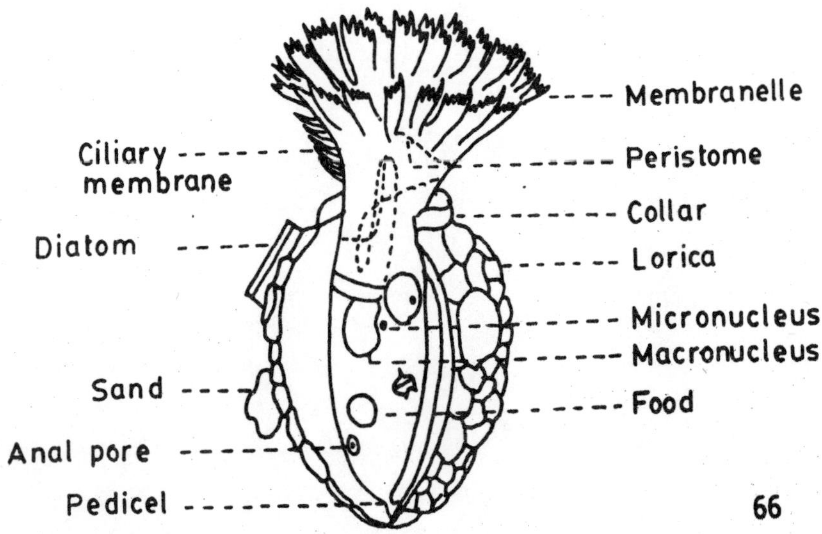

66

Courtesy : Science Reporter

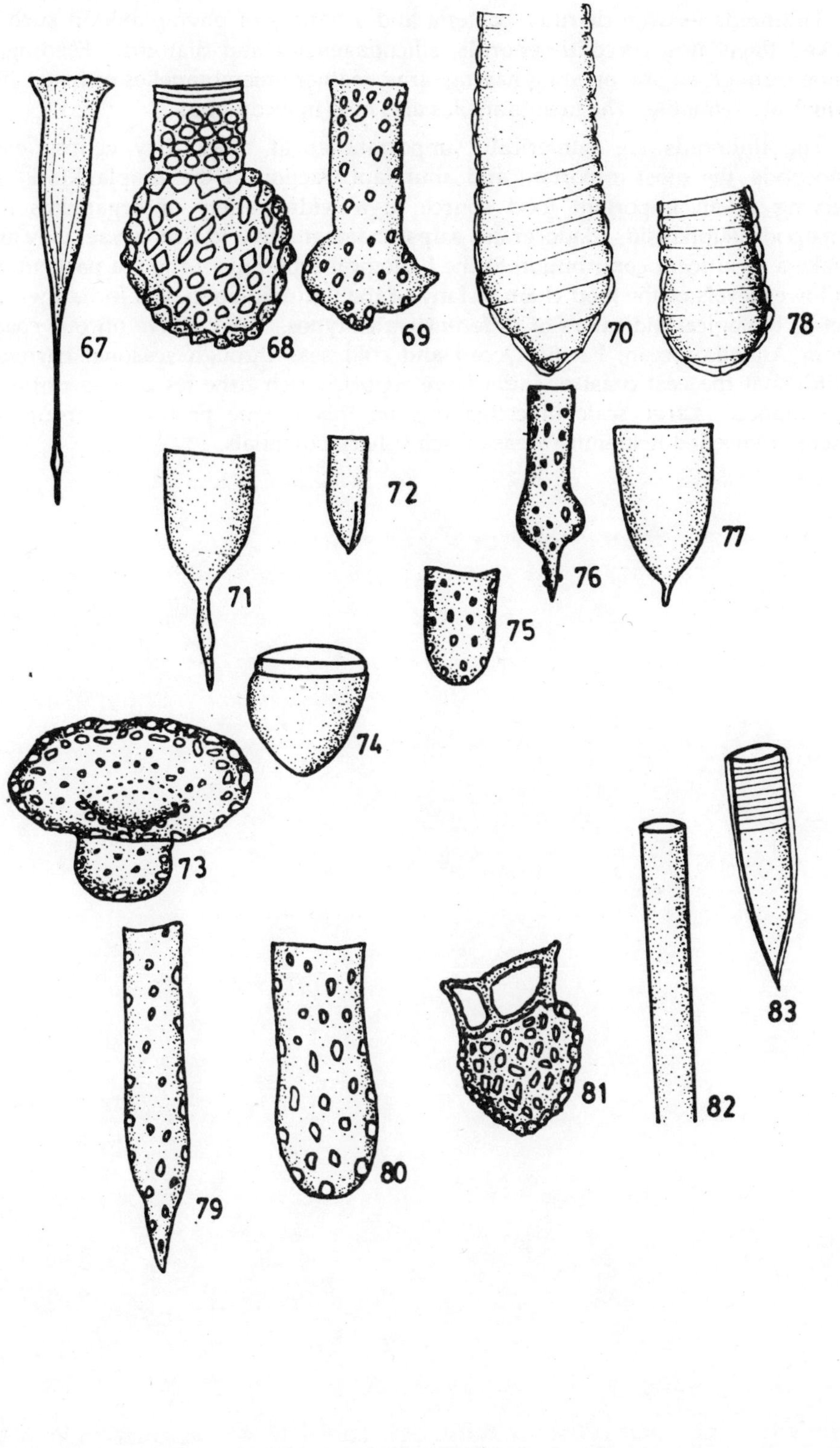
67
68
69
70
78
71
72
76
77
75
74
73
83
82
81
80
79

Tintinnids feed on detritus, bacteria and a variety of phytoplankton such as naked flagellates, coccolithophorids, silicoflagellates and diatoms. Feeding is done through an oral opening having large, feathery membranelles of fused cilia which are retractile. The membranelles also help in locomotion.

The tintinnids are numerically important as at times they eclipse even copepods, the most important and abundant fraction of the zooplankton. By serving as an important food source to a wide variety of organisms like copepods, euphausids, cladocerans, salps, chaetognaths and fish larvae, they also make a significant contribution to the biological economy of the sea particularly at lower levels of the food chain. Many of the tintinnid species of Indian waters act as biological indicators of different water types. They arrive on our coasts from Atlantic Ocean, Pacific Ocean and cold seas through seasonal currents. Studies at the east coast of India have reported rich fisheries due to tintinnid abundance. Large scale investigations on this unique protozoan group are therefore needed to identify areas of rich fishery potentials.

32

Rotifers as Indicators of Water Quality and Pollution

The rotifer fauna (Phylum : Nemathelminthes; Subphylum; Trochalminthes) of freshwaters are rich and varied. Recent ecological investigations have indicated that they are apparently the most sensitive indicators of water quality. As the different species of rotifers tolerate different ranges of oxygen, chloride, pH, alkalinity, temperature, bacterial load and pollution they could be used as reference of the physical and chemical characteristics of freshwater farming systems such as fish ponds, manmade lakes, reservoirs and irrigational canals. Further, many species of rotifers could indicate the nature of fertility and pollution in freshwaters besides helping in the treatment of wastewater systems.

Rotifers as Indicators of Fertility and Pollution

(i) *Fertility* : The presence of *Brachionus angularia* (Fig. 84), *B. calciflorus* (Fig. 85), *B. rubens* (Fig. 86) and *Filina longiseta* (Fig. 87) indicates waters of rich fertility (eutrophic waters.). The associated abundance of bloom forming species of the bluegreen alga *Microcystis aeruginosa* with the above said species corroborate this. In aquaculture systems, the presence of eutrophic species avoids the applications of supplementary feed for the growing fishes.

(ii) *Indicators of Pollution: Rotatoria neptunia* (Fig. 88) *R. rotatoria* (Fig. 89) and *B. calciflorus* are polysaprobic and are capable of tolerating heavily polluted waters including septic tank waters. On the other hand, *B. urceolaris* (Fig. 90) indicates moderately polluted waters and *Filina longisetu and Keratella cochlearis* (Fig. 91) are present only in clean waters or in waters of very slight pollution.

Rotifers in Waste Water Treatment: As the organic matter decomposing bacteria form one of the major food items of rotifers, the latter could efficiently bring down the bacterial load of polluted waters. Further, many species of rotifers feed on the suspended solids and colloids of wastewaters thereby, these species help directly in the biological treatment of waste water purification plants such as biofilters, activated sludge and stabilisation ponds.

Rotifers as Indicators of Water Quality

(i) *Oxygen* : Most of the rotifers possess no respiratory organs and respire by their whole body surface. For this reason, they are unable to live in an anaerobic milieu. *R. neptunia* and *R. rotatoria* tolerate however microaerobic habitats. The presence of such species therefore indicates poor dissolved oxygen content of freshwater systems especially in fish ponds which necessitates the aeration of the latter to avoid large scale mortality of the growing fish.

Courtesy : Science Reporter

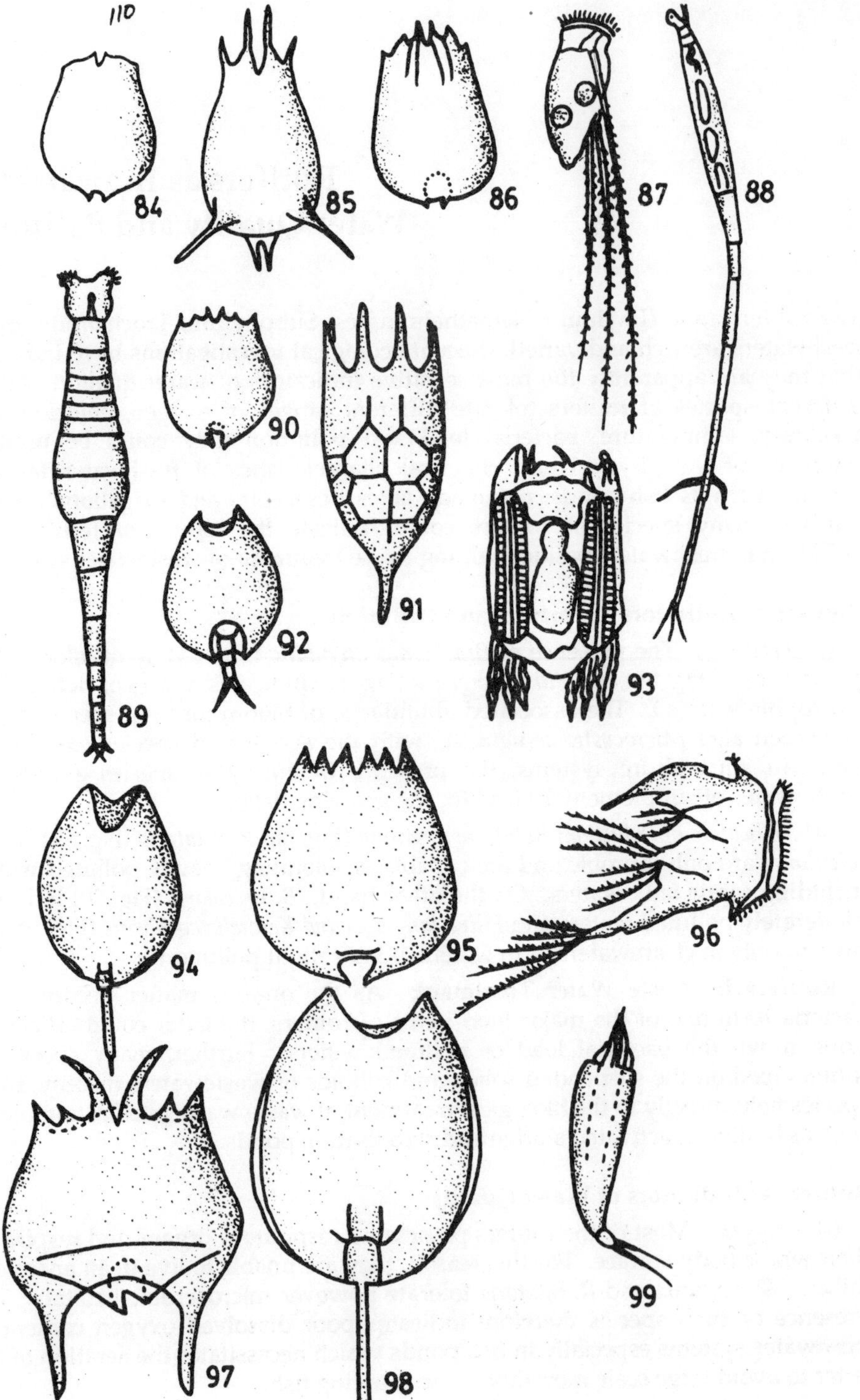
84
85
86
87
88
89
90
91
92
93
94
95
96
97
98
99

(ii) *Chloride*: Increase of chlorinity in fish farming systems affects the physiology of growing fish especially their osmoregulation. Rotifers serve as indicators of waters with varying chloride levels. *Lapadella ovalis* (Fig. 92) and *Polyarthra vulgaris* (Fig. 93) tolerate chloride at 0-20 mg/1 and *Lecana bulla* (Fig. 94) occurs in waters containing chloride at 25-60 mg/1. Interestingly *B. plicatilis* (Fig. 95) and *Hexarthra fennica* (Fig. 96) indicate the brackish nature of their habitat as they tolerate high chloride levels (> 1g/1).

(iii) *pH*: Based on the occurrence of rotifer species, the aquaculture system could be designated as alkaline or acidic. While species like *B. plicatilis, B. calciflorus, B. quadridentatus* (Fig. 97) and *H. fennica* occur mostly in waters of >pH 7, *K. cochlearis, L. bulla and Lecane lunaris* (Fig. 98) are mostly restricted to waters of <pH 7.

(iv) *Alkalinity* : *P. vulgaris* a ubiquitous species occuring only in very low alkaline conditions (0-60 mg/1). *B. plicatilis* is reported to be an indicator of pollution in brackishwaters with low alkalinity. In contrast *B. calciflorus* has wide akalinity tolerance (100-600 mg/1).

(v) *Temperature: Lecane lunaris* and *Trichocerca similis* (Fig. 99) are eurythermal and tolerate wide variations in temperature (16-24°C). On the other hand, *B. plicatilis, B. urceolaris* and *H. fennica* are warm stenothermal (> 24°C) and some species of rotifers are found to be cold stenothermal (< 16°C) indicators.

33

Planktonic Molluscs as Indicators of Ocean Currents

Molluscs, the class of creatures with shells, are generally benthic animals, that is, they live on the sea bottom. But a few are specially modified to lead a holoplanktonic life floating on water. Such animals are generally abundant in the warmer parts of the ocean. It is interesting to know that some planktonic molluscs such as the species of *Janthina, Limacina* and *Clione* periodically drift in vast swarms into the North Sea from the Atlantic Ocean due to water currents and winds. As a result, they serve as useful indicators of movements of water-masses and associated fishery resources in these areas. Further, many of these molluscs living in the open sea like foraminiferans and radiolarians form extensive deposits (oozes) on the sea bottom. Although considerable information is available on the benthic molluscs, knowledge about the planktonic molluscs is scanty.

Janthina (Fig. 100) (Gastropoda: Mesogastropoda), an eyeless, violet snail, has an exceedingly thin and transparent lavender blue shell. It has a very strong bubble float (raft) formed by the secretion of viscous mucus. The animal clings to this raft and hangs upside-down on the surface of the sea. A long row of egg capsules may also be seen beneath the bubble raft occasionally. *Janthina* feeds on the polyps of coelenterates by immobilising their stinging cells through the secretion of violet dye. Interestingly, its raft may also harbour small crabs.

100

The pteropods (those possessing wing-like structures) belonging to the opisthobranch molluscs are fast swimmers and are of two types: those with a shell and those without a shell. The front part of the body of these animals is drawn put on either side into two symmetrical wing-like lobes and hence they are popularly known as 'sea butterflies'. These lobes or parapodia are the

Courtesy : Science Reporter

modified foot and form an important part of the feeding mechanism. *Limacina* (Fig. 101), a primitive pteropod possesses a spiral shell, the diameter of which is only 30 mm. *Creseis* (Fig. 102), an advanced mollusc, has a straight and tapering shell. *Cavolinia* (Fig. 103) has a transparent uncoiled shell which is in the form of quiver or vase. The maximum length of these shells may be a few centimetres. The lobes of these shelled and herbivorous pteropods are ciliated and, by the action of the cilia, food particles such as diatoms and other small organisms are sent to the mouth in a steady stream. Further, by the slow flapping movement of these lobes, the animals progress through water.

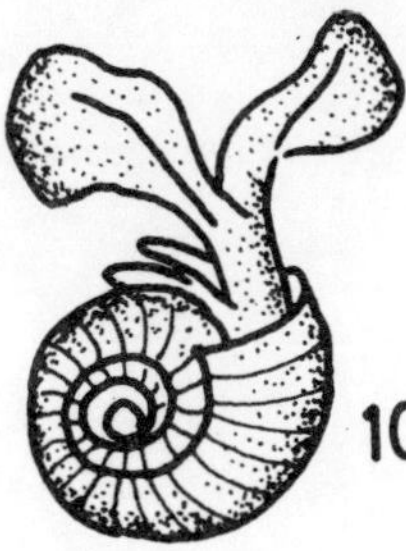

101

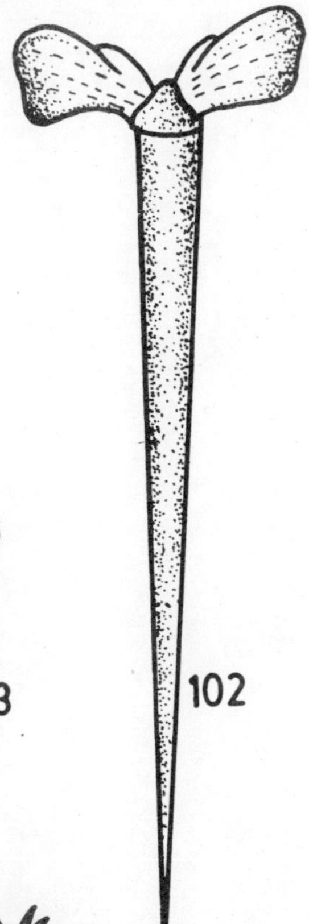

102

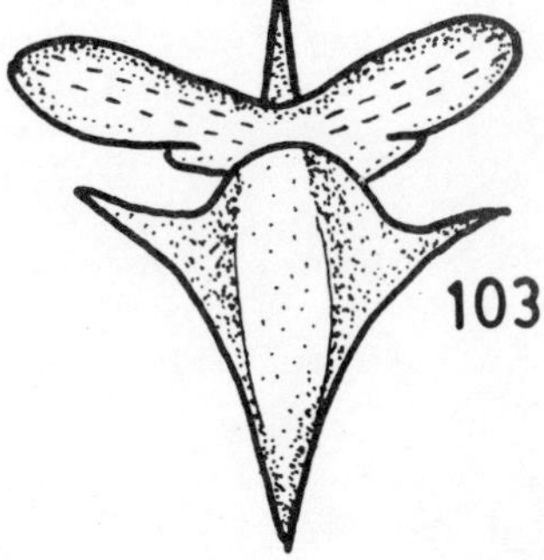

103

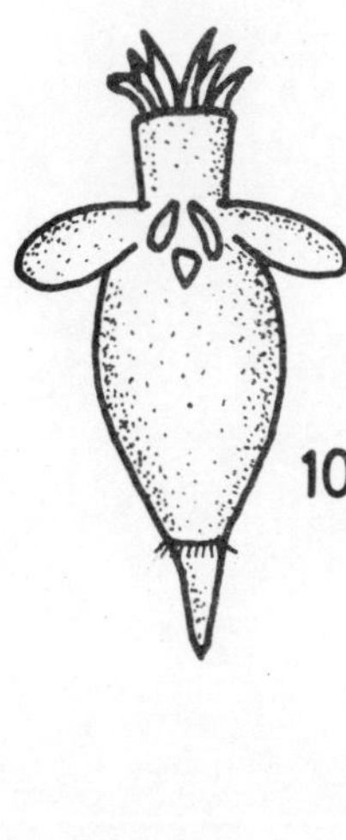

104

The naked sea butterflies do not possess shells, mantle cavities or gills. Their lobes, which are attached ventrally behind the head, produce effective strokes in the dorsal ventral plane. *Clione* (Fig. 104), a naked carnivorous animal, measures only one centimetre and feeds on its shelled cousins. For the capture of prey, there are complex tentacles possessing suckers with hooks. These tentacles can contract into the mouth cavity when the animal is resting.

34

Menace of the Marine Foulers and Borers

The bottom of a ship or boat, painted with an ordinary paint, soon becomes home for a variety of marine foulers, viz., barnacles (Fig. 107), moss animals, tube worms, clams and sea-squirts and wood borers, viz. crustaceans, *Limnoria* (Fig. 108), *Sphaeroma* (Fig. 109) and molluscan shipworms (Fig. 110). They are mostly confined to warm waters of high salt content. The foulers, particularly barnacles, attach in large numbers to the outer surface of the ship and seriously slow down speed of the vessel and, thereby increase fuel consumption. The borers, on the other hand, take shelter inside the fixed and floating timbers and feed on for their nourishment on cellulose. Consequently, the timber is reduced to soft saw dust. It is said that Columbus lost all his ships during his IV voyage as a result of the ravages of shipworms.

How do these Organisms take Shelter

The larvae of barnacles (Fig. 105, 106) swim actively for a number of days to find some suitable surface to land on. When a young barnacle finds such a surface, it crawls around over a small area in search for a choice location. It, then, anchors itself firmly on its two front legs with the help of a cementing material and slowly begins to form its familiar conical shell.

The borers, particularly the shipworms, release their larvae which are about .025mm thick. After 24 to 48 hours of free swimming, they bore and make a hole in the wood which resembles a pin point. Once they get into the wood, they grow rapidly and dig burrows which may be as much as 1.25 cm in diameter and 30 cm in length. The larvae may survive even when the wood is kept out of water for more than two weeks. It is, however, interesting to observe that fouling and boring do not occur simultaneously. Dense fouling effectively inhibits attack of shipworm borers as adult barnacles feed on shipworm larvae.

Measures for Protection

In India alone, the periodic cost of replacement of fishing crafts destroyed by marine borers amounts to Rs. 2.5 million. In the United States, damage has been found to exceed 50 million dollars every year. With a view to eradicate foulers and borers antifouling and boring paints have been developed nowadays. Most of them contain cuprous oxide which dissolves slowly in seawater and is very poisonous to fouling organisms.

There are basically two different types of paint containing cuprous oxide. One is the 'continuous contact' type, which has considerable amount of the poison in

---

Courtesy : Science Reporter

105

106

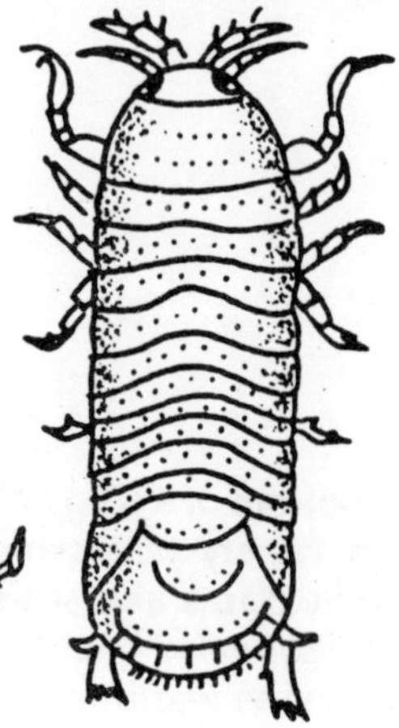

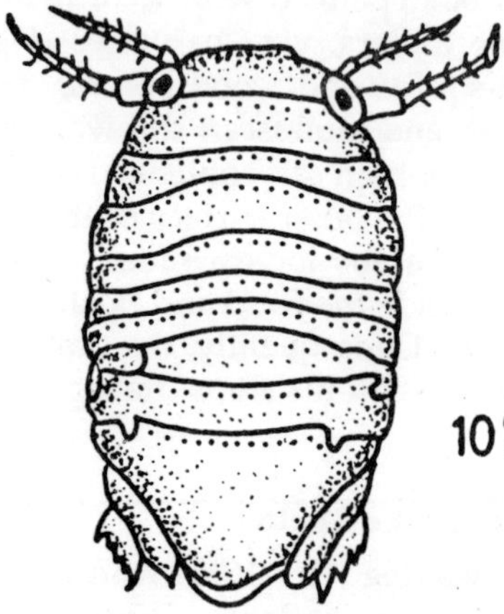

107

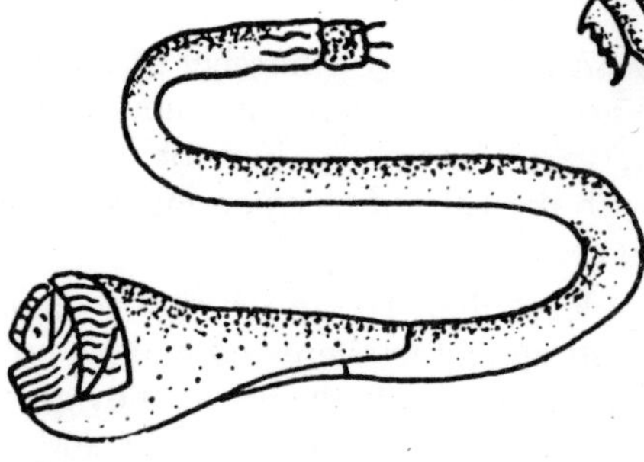

108

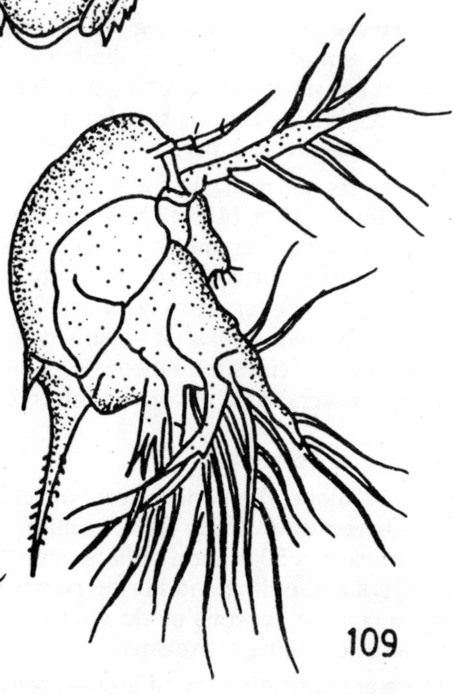

109

110

the surface layers of the paint. Such paints need to be applied as a thin coat. The second type, known as 'soluble matrix' paints, contain a lower concentration of the poison but have a porous structure which allows poison to reach the surface from within the paint. The latter type is applied as a thick coating. The soluble matrix paints are more effective but both the types resist fouling only for two years. It is always advisable to have a prepainting with a special paint before these paints are applied because they may cause considerable corrosion. Furthermore, vessels painted by these paints should be immersed in seawater before they are dry. Otherwise, the surface of the paint hardens and seals the poison beneath making it ineffective.

Scientists have recently found that a group of organic compounds of tin could also be equally effective like cuprous oxide paints. Two compounds, one liquid, technically known as bis (tri-n-butyltin) oxide and commonly called TBTO, and the other, solid, tri-n-butyltin fluoride (TBTF), have been successfully used in antifouling paints. These modern paints have several advantages. They are not corrosive. Also, they do not lose their effectiveness on drying. Consequently, boats do not need to be launched immediately after painting and may be repeatedly hauled without destroying the properties of the paint. Another method of controlling foulers and borers is the periodic taking of vessels to low saline regions like rivers which is said to kill these organisms.

35

Beautiful Corals

Most of us are aware of corals but perhaps are not familiar with their biology. For a layman corals are mere inanimate calcareous matter of intricate design and beauty. Corals are actually the skeletal remains primarily of calcium carbonate secreted by living coral polyps (colonial animals) belonging to the phylum Coelenterata.

Formation of Coral Reefs

Minute larval forms of corals given off in myriads by adult forms float passively in ocean waters; if they happen to drift with currents of suitable temperature to a shore of firm rock or to a shallow bank where pebbles or shells lie undisturbed by surface waves they may attach themselves there and in time form a new reef. A coral reef thus is a result of the activity of millions of coral polyps over several thousand years. Though coral reefs are seen in depths of 10 m to 50 m, most abundant beds occur in shallow seas only. Coral reef surveys reveal that they are confined to tropical and subtropical seas and are scattered over an area of 190 million square kilometres as a belt around the globe on either side of the equator between latitudes 35° N and 32° S.

Reef Types

Coral reefs are mainly of 3 types, viz. Fringing reefs, Barrier reefs and Atolls. By a process of evolution fringing reefs become barrier reefs later giving rise to atolls (Fig. 111, 112, 113). In seas around India, all these types are present. 'Fringing reefs' form shallow shelves in shallow waters at or near the shore of the main land or around offshore islands (Fig. 111). Heavy sedimentation and freshwater runoff along many mainland coasts make these less attractive. At the southern end of Indian peninsula, this type is seen as a chain of well developed reefs starting from Rameswaram Island and extending beyond Tuticorin. They are also present in the Gulf of Kutch on the West Coast of India. Barrier reefs are long, ribbon like reefs and are asymmetrical in plan and section (Fig. 112). The famous Great Barrier Reef of North-eastern Australia which is 900 nautical miles long is the largest of its kind. Atolls are ring shaped reefs enclosing lagoons. The Laccadives and Maldives possess both barrier reefs and atolls.

Biology of Coral Animals

Though majority of coral animals are colonial in nature (Fig. 114), a few of them lead solitary life. Each coral is a tubular unit. Its epidermis is modified to form the so-called calico-blastic layer and secretes a cup-like skeleton of calcium carbonate composed of calcareous fibres radiating from a central core probably of

Courtesy : Science Reporter

Fringing reef

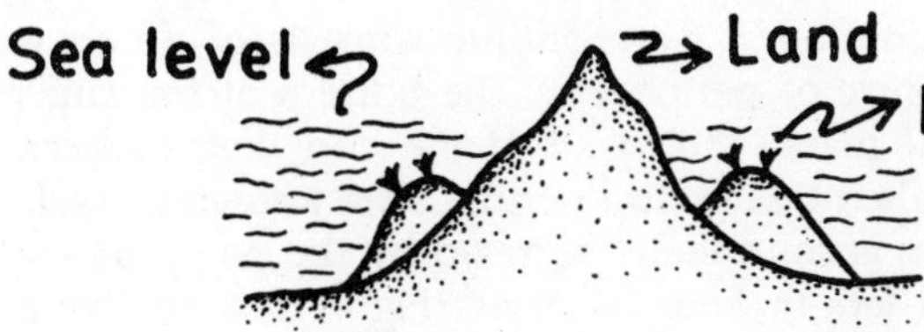

111

Barrier reef

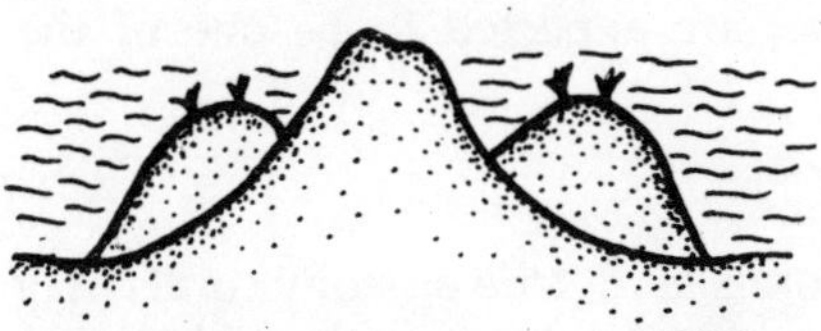

112

Atoll

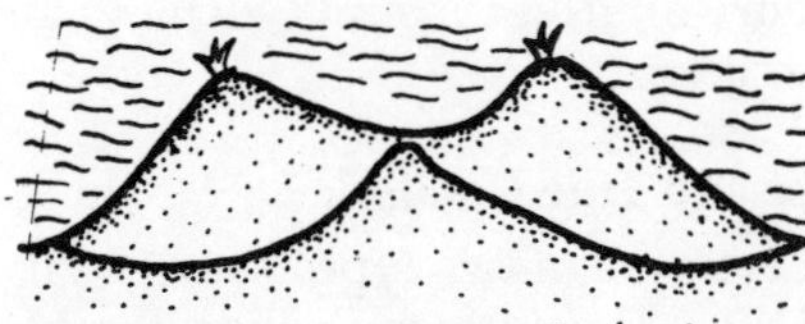

113

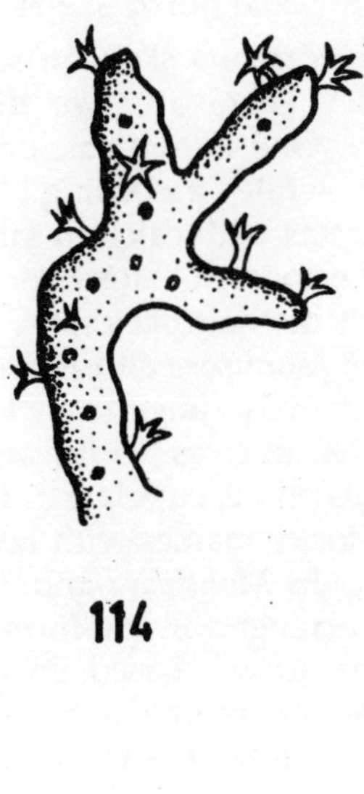

114

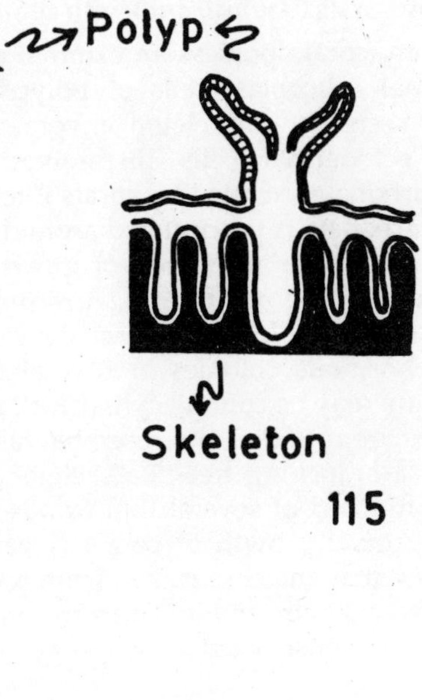

115

organic material. A considerable part of each coral polyp is inseparably fixed in this cup-like skeleton. The skeleton of a colony as a whole is termed as 'Corallum' and of each polyp as 'Corallite'. The corallite consists of a cup with vertical ridges radiating from centre of periphery. The bottom of the cup beneath the polyp is called the basal plate. The wall of the cup that encloses the aboral portion of the polyp is called theca and the ridges are known as skeletal septa or sclerosepta. The skeleton is continually secreted by the polyp bases so that, the corallum grows both in length and diameter (Fig. 115). In living condition, polyps protrude outside from their cavities with their tentacles extended. They fully extend especially during night when they feed on zooplankton. At a depth of a metre or so polyps protrude 4 cm or 5 cm above the level of corallum in bright sunlight. The corallum is thus completely concealed from view. Corals are highly coloured organisms. Blue, pink, liliac, violet, eosine red or tan are very common. Various parts of colony may have different colours. They are also known as 'sea flowers' because of lovely colours. Live corals are a treat to the human eye and are nature's marvels. In the proposed 'Marine Park' to be set up in our 'coral sea', corals, if they survive, are expected to be one of the star attractions.

False and True Corals

Corals are of two types namely 'false corals' and 'true or stony corals' (Fig. 116 & 117). The former include the blue coral (*Heliopora*) with a blue skeleton, gorgonaceans such as sea fans, sea whips and sea feathers with a skeleton in the form of scattered horny or calcareous spicules or a solid axial structure and the black or thorny corals (*Antipatharia*) with central axial horny skeleton.

True or stony corals possess an external calcareous skeleton secreted by the outer epidermal calicoblast cells of polyps. There are two different views regarding the secretion of skeleton in corals. One is that calcium carbonate is secreted by ectodermal cells themselves, while according to the other, ammonium carbonate excreted by corals interacts with calcium salts of seawater and calcium carbonate is precipitated around ectoderm. Stony corals show wide variations in form, size and mode of growth of their skeletons. *Fungia* has a single corallite. On the other hand, *Acropora, Montipora* and *Porites* are colonial with several corallites. In size a coral skeleton may range from 1 mm to 2 mm diameter to enormous colonies of several centimetres width and height. A solitary corallum may be conical, cylindrical, patellate, cupolate or fan shaped. In *Porites* coallites are rounded and resemble hillocky masses with heavy skeletons. Acropora has a branching tree-like skeleton. In *Montipora* and *Echinopora,* the corallum is composed of several thin fronds arranged in the form of petals in a flower. The annual growth of corals is very slow. Based on environmental conditions, they may increase in size from few millimetres to 5 cm. Fresh corals have grown over 30 cm within Tuticorin harbour where the breakwaters have only recently been constructed.

Economic Importance

Some corals are highly priced for their decorative value. *Corallum rubrum* is considered as a 'precious stone' in the orient. As it was believed to have sacred

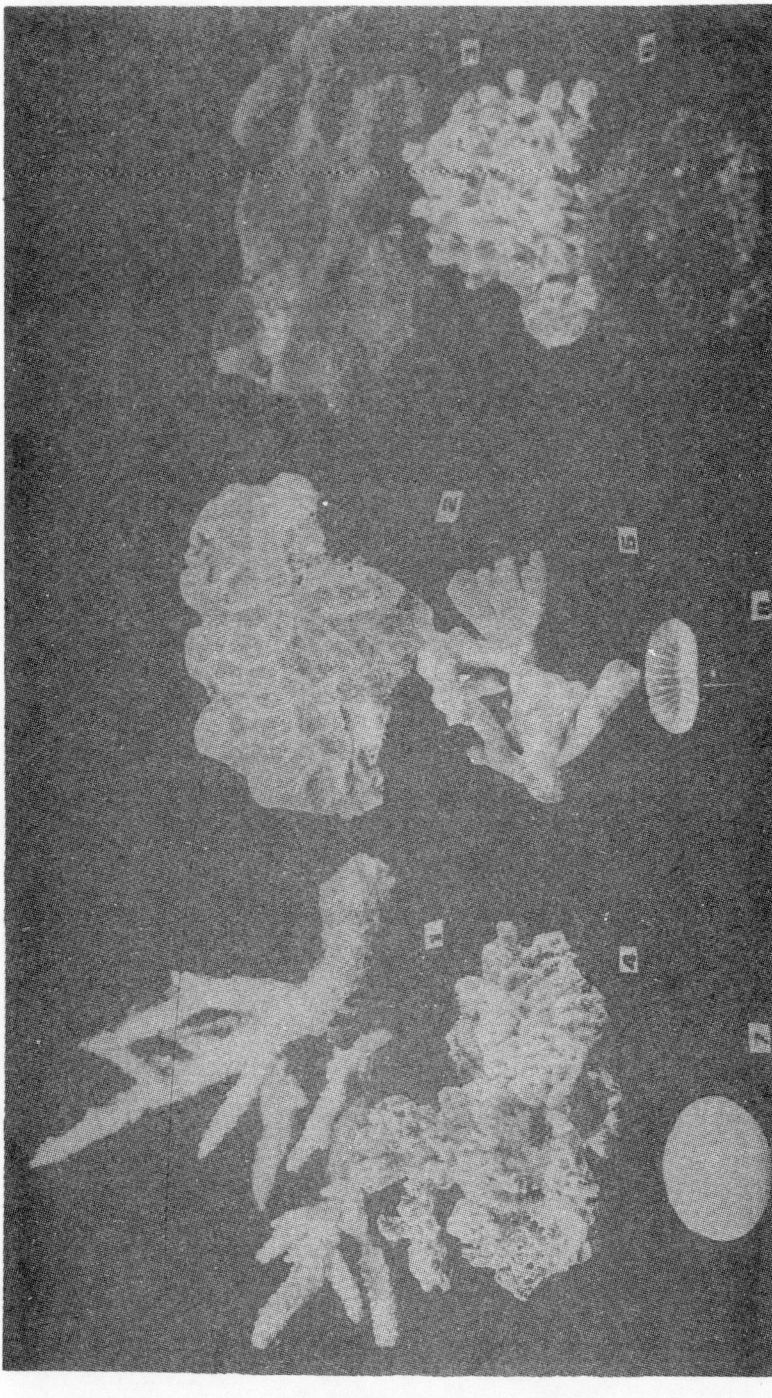

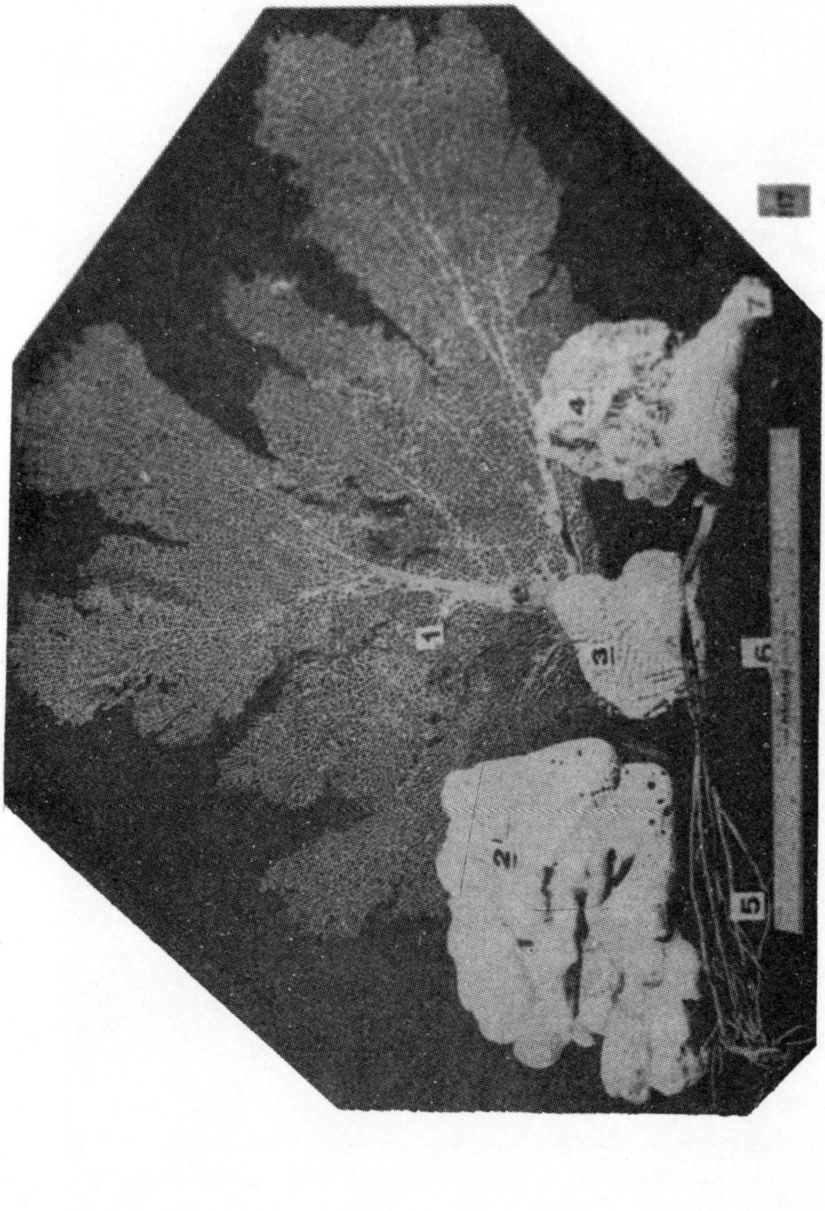

properties, people of India and China were said to have exchanged this type of coral for emeralds, rubies and pearls since ancient times. The Romans used to hang branches of coral around necks of their children to safeguard them from danger. Italians are said to wear this to ward off evil eye and also as a cure for female sterility. The red coral is used in medicine since ancient times and even now the organ pipe coral (*Tubipora*) has a place in some indigenous systems of medicine in South India. People of Malay Archipelago and Japan wear ornaments of black corals as preventives for rheumatism.

Chunks of coral skeletons especially of species like *Porites* spp. are used as building blocks and for metalling roads (Fig. 114). As they are porous, they are said to make houses cool. Corals serve as raw material for the preparation of lime, mortar and cement because of their calcium carbonate content. Some older coral lime stones are rich in magnesium (dolomotized), hence their great value in making cement. They are also used in the manufacture of carbide in South India. Coral skeletons help formation of Islands and may act as natural barriers against sea erosion and cyclonic storms. Old reefs have long served as sea bird sanctuaries and have got bird droppings (guano). When such islands have been long elevated above sea level, the guano has leached downward into the coral lime stones leading to phosphatisation, hence the name phosphate islands (e.g., Nauru, Ocean Islands, Christmas Island) which are valuable sources of industrial phosphate. Coral polyps lodge in their tissues some photosynthetic algae (zooxanthellae) in a symbiotic association and benefit by the extra oxygen produced by photosynthesis. Very often it leads to supersaturation of oxygen in the water around the reefs which is the cause of high productivity of the coral sea.

Coral reefs favour rich submarine organic growth and show a good association of fish varieties. They also serve as good nursery grounds for commercially important fish. Reef fish varieties are more colourful than palatable. Octopi, sea urchins, sea slugs, trochus, oysters and turtles are the other valuable fauna present in coral reefs.

Coral reefs can also cause difficulties to men; ships have been wrecked on unmarked reefs. There are still broad areas in the Great Barrier Reefs of Australia, Coral sea and South China sea marked on the hydrographic charts with the warning 'dangerous ground-numerous uncharted reefs'.

The indiscriminate exploitation of coral reefs on our coasts for making lime and cement is changing their ecology markedly. Damages to the rich fauna and flora of the reefs and surrounding areas are obvious. Protests and legislation have not helped the situation and permanent damage has occurred to the beauty and peace of our unique coral sea. Live coral polyps are often attacked and eaten by animals like starfish and some fishes. Certain parts of the world have atttempted to eradicate them and to preserve the natural beauty of corals. Malaysia and Australia have tried to set up artificial coral reefs. Unused rubber tyres tied in a pyramid form in the sea attract coral polyps settlement and an assemblage of fauna and flora. Can we rehabilitate our coral reefs this way? Much more ingenuity and love for one of nature's marvels would be required for the purpose.

36

Oceanic Insects

Of 1.5 million species of insects known so far, about 30,000 species are found in aquatic environments. Of this, only five species in all belong to the genus *Halobates* which are restricted to open ocean. So few species of insects present in oceanic environment could be attributed to low concentrations of calcium and high surface tension of seawater.

Oceanic insects have adapted themselves to environment, for they have no wings and breathing siphons. The various species found in oceans are *H. micans, H. germanus, H. sericeus, H. sobrinus and H. splendens,* of which the first two are found in Indian waters. The members of the genus *Halobates* are commonly called ocean skaters and are found on the surface of oceans. They belong to the family Hemiptera of the class Insecta. Individuals are dull coloured or greyish and measure about 3.5 mm to 6.5 mm in length. The general structure of water skaters is different from that of common insects in many ways which enables them to survive on the surface of water. If submerged in water, the oceanic insects carry with them a large bubble of gas entangled in the hair coat of their body and appendages, which makes them buoyant. The hairs present on their surface have also a waxy layer which makes them waterproof. If exposed to water for prolonged periods, the hairs get wet and so the insects have to dry their body.

Ocean skaters feed upon dead remains of medusae, sea anemones, crustaceans and fish eggs and larvae. Their enemies include birds, particularly phoenix petrel, and surface-feeding fish. The eggs of *Halobates* are 1mm-1.2mm in size. Each gravid female carries about 40 eggs and usually several females clump together during the laying of eggs. (Fig. 118-120)

Courtesy : Science Reporter

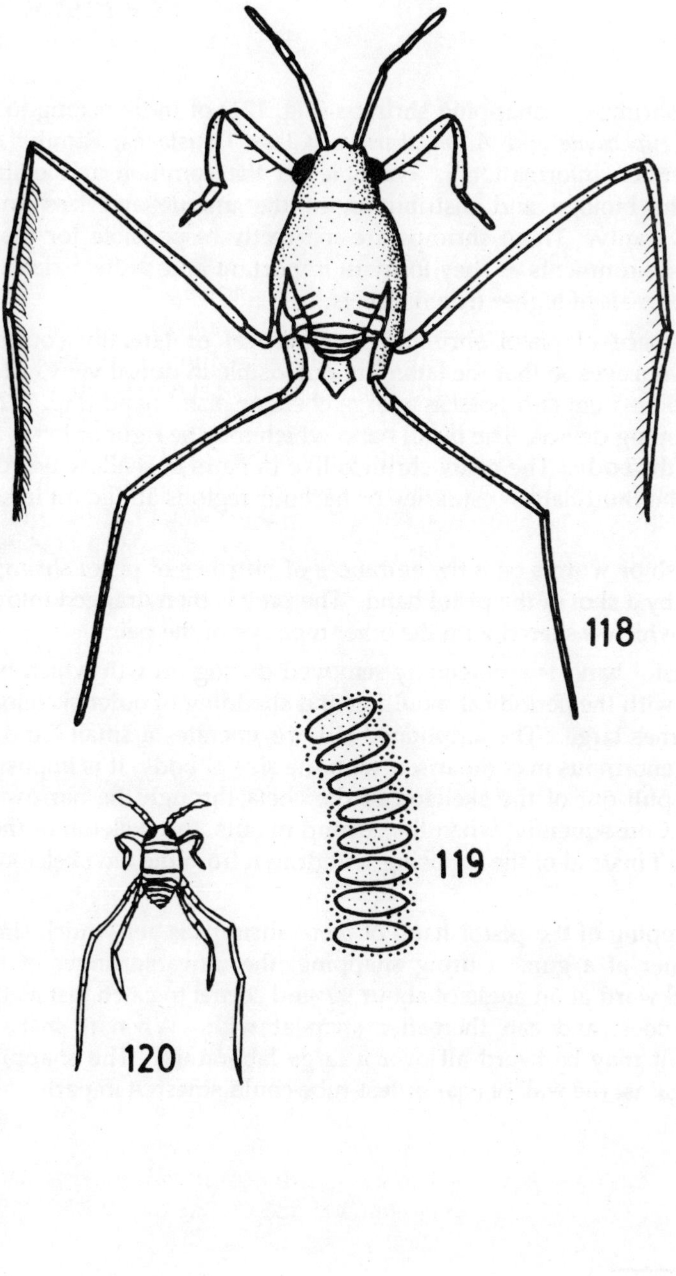

118

119

120

37

The Pistol Shrimps

The pistol shrimps or snapping shrimps (Fig. 121) of India belong to the species of *Alpheus euphosyne* and *A. malabaricus* (Class: Crustacea; Family: Alpheidae). Although much information is available on the common and edible penaeid shrimps, the biology and distribution of the unique and less known pistol shrimps is scanty. These shrimps are indirectly responsible for the fertility of intertidal environments as they form an important and protein-rich food source for many animals of higher trophic levels.

The carapace of pistol shrimps is cylindrical or laterally compressed and usually covers eyes so that the latter are not visible in dorsal view. They grow to a size of 2.5cm-5 cm and possess a large chela or pistol hand (Fig. 122) equipped with a snapping device. The pistol hand which may be right or left is at least half the size of the body. The pistol shrimps live in pairs in shallow burrows formed mostly in the mud flats of estuarine or harbour regions abundant in seaweeds or shells.

When fish or worms pass the entrances of burrows of pistol shrimps, they are paralysed by a shot of the pistol hand. The prey is then dragged into the burrow for a meal which is shared with the other member of the pair.

If the pistol hand is accidentally removed during growth which is invariably associated with the periodical moulting and shedding of outer skeleton, the other hand becomes large. The amputated side regenerates a small hand. Since the chela is so enormous in comparison with the size of body, it is impossible for the shrimp to pull out of the skeleton of the chela through the narrow leg during moulting. Consequently, when the shrimp moults, the skeleton of the chela and leg flakes off instead of the leg being withdrawn from the old skeleton as in other decapods.

The snapping of the pistol hand of these shrimps is very much similar to that of a hammer of a gun. During snapping, the movable finger of the chela is turned backward at an angle of about 90° and seems to catch just as the hammer of a gun does, and can thereafter snap at will. When it snaps within an aquarium, it may be heard all over a large laboratory. The snapping of these shrimps against the wall of a jar or test-tube could smash it in parts.

Courtesy : Science Reporter

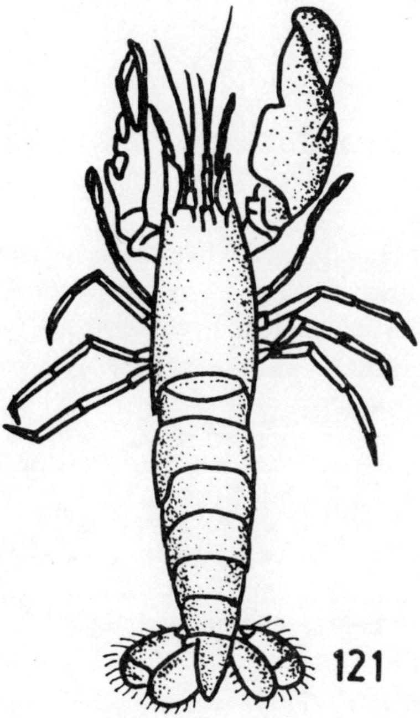

121

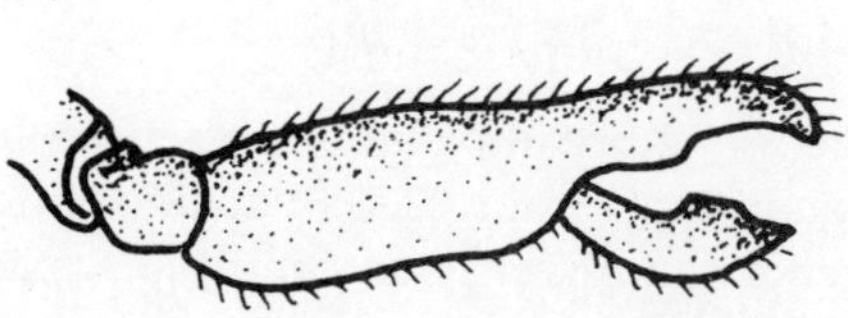

122

38

Sea Cucumbers are Rich in Protein

The sea cucumbers or holothurians (Fig. 123) belong to the invertebrate Phylum, Echinodermata. They are confined to marine environment and are inhabiting all seas and in depths from the intertidal zone to even depths of more than 10,000 m.

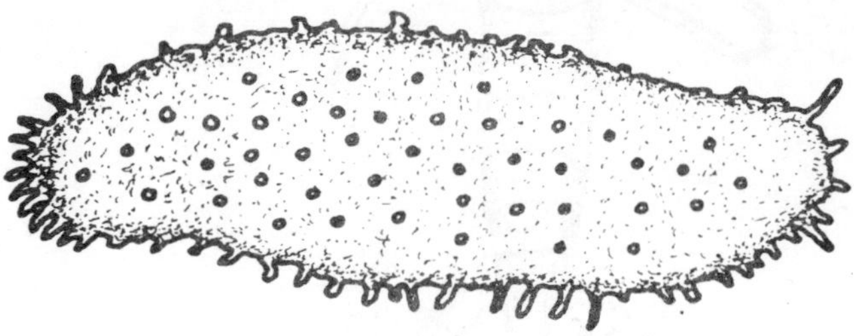

123

In the intertidal zone they live wholly or partly buried in muddy and sandy bottom. Most of them are dark to brightly coloured.

Sea cucumbers feed on decaying organic matter present in mud. In polluted waters and in waters of higher temperatures, sea cucumbers send their viscera out. The viscera can be regenerated under good conditions. Many holothurians are nocturnal in habit.

There are about 50 species of holothurians and species such as *Thelonota ananas, Holothuria nobilis, H. mauritiana, H. scabra, H. lecanova, H. argus, H. edulis* and *H. echinitis* are of great economic importance.

The processed body wall of holothurians is known as 'beche-de-mer' or 'trepang'. For obtaining beche-de-mer, sea cucumbers are first boiled to make them eviscerate and to shorten and thicken. Then the body wall is dried in sun or by smoking or dried by combinations of both.

The protein content of beche-de-mer varies from 35 to 65 per cent. Minerals like calcium, sodium, potassium and iron along with protein increase the nutritional value of beche-de-mer.

While using beche-de-mer, dried body wall is cut into small pieces, which are added to soups and stews and are said to impart a delicate flavour. The cooked pieces swell into a gelatinous condition and are also eaten as titbits.

Courtesy : Indian Express

In some parts of Indo-Pacific, holothurians are irritated until they send out all their internal organs, including cuverian tubules and gonads and are then eaten raw.

Considering the protein content, it is reasonable to believe that beche-de-mer could do spectacular things for our stamina. Furthermore, it is highly digestible as the protein is totally soluble in the enzyme pepsin. It is also used now for treatment of high blood pressure.

Beche-de-mer is a good foreign exchange earner. Large quantities are exported to Hong-Kong, Malaysia, Singapore, Taiwan, Japan, U. K. and U.S.A.

As per the report of the Marine Products Export Development Authority, 91.3 tons valued at Rs. 2.1 million have been exported form India in 1975. To meet the demand, culture of *H. scabra* is being attempted by Central Marine Fisheries Research Institute at Aberdeen Jetty in Port Blair.

Though many species of sea cucumbers are useful to mankind some are toxic. The poison of sea cucumbers is a chemical 'holothurin' which is concentrated in the 'organs of Cuvier'. This poison seems to have nerve blocking effect in laboratory animals. But its effects on humans have not been fully determined.

However, in humans, symptoms such as burning pain, redness and a violent inflammatory reaction due to dermal contact with poison of sea-cucumbers have been reported. If the fluid contacts the eyes of victims, blindness may result. Ingestion of poison may be fatal too.

But recent studies suggest that holothurin could be used medically like snake venom. To cite an example it is now known to suppress the growth of tumour cells - an antimicrobial property.

39

Why Should We Eat Fish ?

The importance of fish and its farming is understood through an old Chinese proverb 'Give a man a fish he lives for the day; teach him to fish and he will live for his whole life.'

Importance of Fish in Human Nutrition

Fish in one form or other forms is an important and indispensable item in human diet since prehistoric times. Fish and its products are highly nutritious and are excellent sources for dietary essentials such as protein, minerals and vitamins. The energy content of fish is more (193 K. cal/100 gms) than that of beef (114 K. cal/100 gms) and chick egg (173 K. cal/100 gms). A serving of 100 grams of fish every day could satisfy half the total protein required by the body. About 90 to 100% of fish protein is digestible. Hence, fish foods are included in many special diets for people with digestive disorders and for convalescent ulcer patients. Furthermore, due to the unique chemical nature of fish fats, fish foods are often recommended to people who suffer by heart disease. The vitamin A content of fish livers helps for good vision. Other vitamins B6, B12, Biotin and Niacin and minerals such as phosphorus, potassium and iron help in normal growth of human body. The level of calcium, magnesium and chlorine in fish is also greater than that of meat. Several trace minerals like copper, zinc, manganese, cobalt, molybdenum and selenium are found in fish eggs (roe). Due to rich iodine content of marine fish, fish eating people seldom get goitre. The low incidence of tooth decay in some people has been attributed to high consumption of fish. More recent medical evidence suggests that a high dietary intake of fish which contains fluoride may prevent certain bone diseases.

Percapita Consumption of Fish

The average annual percapita consumption of fish in India is very poor compared to many countries as shown below:

U.K.	...	19 kg
U.S.S.R	...	16 kg
Denmark	...	11 kg
Germany	...	9 kg
France	...	9 kg
U.S.A.	...	7 kg
Italy	...	5 kg
Switzerland	...	3 kg
India	...	2 kg

Courtesy : Kisan World

In India the higher consumption of fish is generally observed in the maritime states of westcoast. Kerala has the highest annual percapita consumption (10 Kg) followed by Bombay (6 kg), Karnataka (3.5 kg), Tamilnadu (2.5 kg), Bihar (1 kg), Delhi (0.5 kg) and Punjab (25 grams).

Preference of Different States

Sharks, skates and rays are popular in southern Bombay and all over Tamilnadu. But these varieties are not having high demand in the main city of Bombay and West Bengal. Catfish and eels are rejected by jews and Shia muslims. Rohu, the freshwater carp and Hilsa, the estuarine fish are preferred in marriage functions in West Bengal. Urban consumers prefer fish with firm flesh with minimum small bones and good taste. For this reason, pomfrets, seerfishes, mullets, bekti, Indian salmon and soles among marine fishes and carps and murrels among freshwater fishes are in great demand.

40

Why Fish Smells ?

A serious problem often met with in intensive fish culture system is the absorption of objectionable odour compounds by the growing fish from the environment. In freshwater culture systems, this odour is usually of an earthy-muddy type. It is caused by a compound called 'geosmin' which is a colourless neutral oil, synthesized by microorganisms such as actinomycetes and blue-green algae, especially *Oscillatoria* spp. Incidence of unpleasant fishy odour is however independent of season, location or density of algal blooms. Some economically important and culturable freshwater fishes such as catla, rohu, mrigal, grass carp, silver carp and common carp, when stocked in confined areas such as in smaller ponds often develop such odours and are at times not preferred by the consumers.

In order to get rid of this objectionable odour, the following methods have been found most suitable.

(i) After harvest fish should be kept in a cement tank containing chlorinated water for a week before they are sent to the market.

(ii) By proper feeding management practices, the accumulation of the feed wastes could be minimised. The growth of geosmin-producing microorganisms is often accelerated by the inorganic and organic nutrients that originate from such feed wastes.

(iii) Periodic and timely change of water in culture systems will avoid the accumulation of unwanted compounds which may otherwise enhance the growth of the said microorganisms.

(iv) The phytoplankton blooms should be eradicated as and when they appear because these blooms are often associated with geosmin-producing blue-green algae.

(v) As clay particles often act as adsorbents for the geosmin compounds, it is necessary to remove clay deposited on the bottom of the tank before the start of fish culture operations.

Trimethylamine oxide (TMAO) is considered to be one of the major compounds responsible for the characteristic odours in marine fishes. This compound occurs naturally in most of the marine fishes to the extent of one per cent of the weight of the flesh. Season, size and age of the fish are some of the factors responsible for the occurrence of this compound in fishes. The TMAO of marine fishes could originate either from protein metabolism as a non-toxic end product or from the marine zooplankton- the chief food organisms of marine fishes. When the fish dies, the TMAO undergoes gradual reduction to trimethylamine, the quantity of which is considered to be a useful index of spoilage in marine fishes.

Courtesy: Science Reporter

41

Puffer—The Most Dangerous Fish

The puffer fish, (Fig.124) otherwise called blowfish, swellfish or evil globefish belong to the genus *Tetraodon* of the family Tetraodontidae. There are about 100 species of pufferfish distributed in almost all tropical and temperate seas. The important characteristic of puffers is their ability to balloon out from a reasonable fish shape into a sphere two or three times larger. That is, when they are freightened, excited or annoyed, they gulp water or air into a sac on the belly. This adaptation achieved mainly by the rib-free pectoral muscles is mainly to discourage predators or intimidate enemies. However, when the fish feels safe, it ejects water or air out thus deflating to its normal shape. It is also interesting to note that the balloon full of water is very strong for even a big man to stand on it.

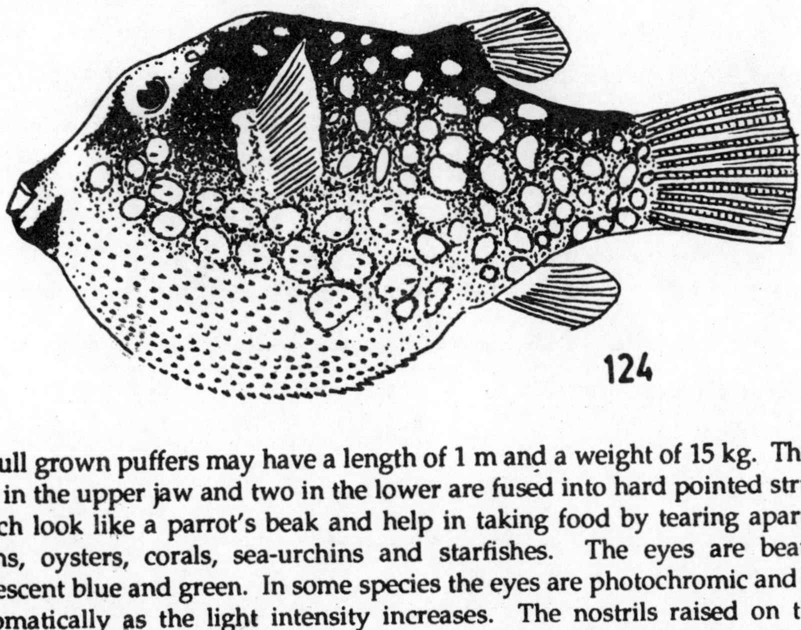

124

Full grown puffers may have a length of 1 m and a weight of 15 kg. The teeth, two in the upper jaw and two in the lower are fused into hard pointed structures which look like a parrot's beak and help in taking food by tearing apart crabs, clams, oysters, corals, sea-urchins and starfishes. The eyes are beautifully iridescent blue and green. In some species the eyes are photochromic and darken automatically as the light intensity increases. The nostrils raised on the tiny tentacles and located below eyes are capable of sensing chemicals in the water. As puffers are without ribs and pelvic fins, they do not swim like other fish and hence, drift slowly undulating their dorsal and anal fins which are floppy and lack both spines and rigidity.

Puffer is considered to be the world's most dangerous fish as its ovaries, intestines, liver, kidneys, skin and eyes contain a neurotoxin called tetradotoxin. This toxin, which has no known antidote, is a nonprotein substance and has a molecular structure unlike any other chemicals previously known to organic chemistry. It is worth mentioning here that the puffer toxin is 275 times deadlier than cyanide. A medium size fish has about 25 mg of toxin and this quantity is enough to kill more than 20 persons, as the lethal dose has been found to be merely one mg. Even a tiny touch of the poisonous organs on the fish flesh makes the latter poisonous. The eaters of contaminated flesh or poisonous organs may die within minutes. The victims before their death experience numbness in legs and arms and respiratory problems.

Although 60% of puffer poisonings have been proved to be fatal, it is the epitome of gourmet dining in Japan. The fibreless, gelatin like puffer meat is carefully cooked and served by well trained and licensed cooks. The puffer toxin owing to its potency, has been found to be an important tool in modern neurological research. In a dilute form it is used as a painkiller for victims of neuralgia, arthritis and rheumatism.

42

Fishes that Shed Skin

Only two species of fishes are known to cast off their skin after the fashion of reptiles. *Congiopodus spinifer* (Family: *Congiopodidae*) collected by Prof. J.L.B. Smith of Grahamstown Institute of Ichthyology, South Africa in 1961 from South African waters was the first species (Fig. 125). This grows to a length of about 35 cm. The exact way in which this species moults its skin is not yet known. In 1968, Dr. V. Ramaiyan of Marine Biological Station, Parangipettai (Porto Novo) collected the second species, namely, *Acanthosphex leurynnis* (Family: Bathyaploctidae) from the trawl net catches of the nearshore waters of Porto Novo, Tamilnadu (Fig. 126). This fish is about 3 cm. in length. When they were kept alive in the Marine Biological Station laboratory aquarium, at Parangipettai, they were found to cast off their skin once in 4 days, and curiously enough, the casting was found to occur around 10 am. The skin of the fish came off in large pieces which contained solid concretions (Fig. 127). The concretions were not soluble either in potassium hydroxide or hydrochloric acid.

Courtesy : Science Reporter

125

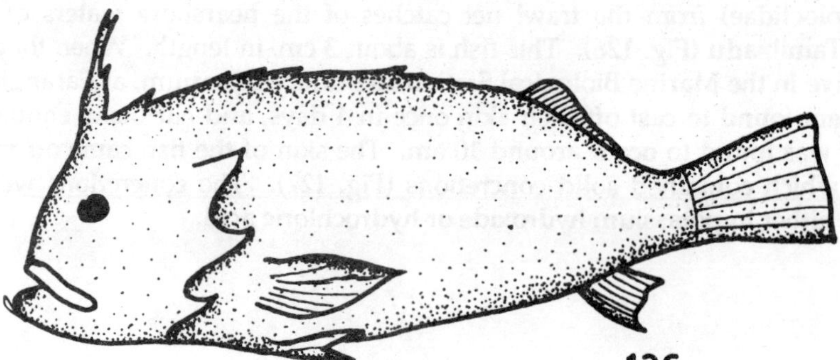

126

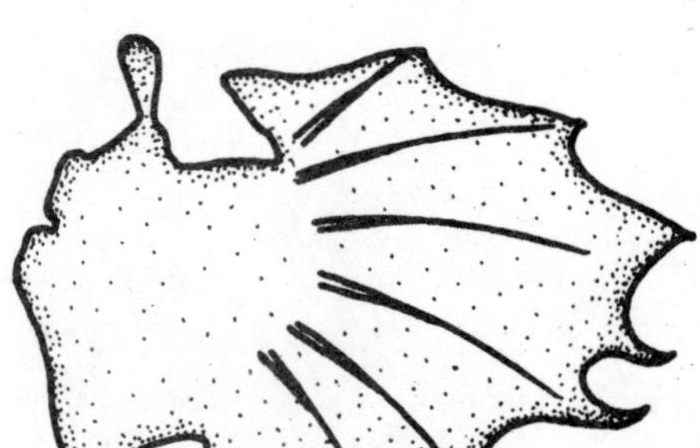

127

43

Unique Devices to Study Fish Behaviour in Polluted Area

Lotic water bodies such as rivers and estuaries and their fish fauna are often threatened by the discharge of effluents from the nearby industries. Different species of fish possess different tolerance limits in these toxic pollutants preferably chemicals and areas of higher concentrations are normally avoided by fish shoals. Laboratory studies on the behaviour of fish in relation to various concentrations of toxic chemicals are therefore imperative. Such studies would help locate not only areas free from pollutants but would also give an idea on the concentrations of toxic elements and the associated size of fish assemblages at the different points of water course. At present, sophisticated instruments are used in the study of fish behaviour.

Blazka Apparatus (Fig. 128, 129): It consists of two concentric cylindrical tubes made of transparent acrylic plastic material. The inner tube is fixed inside the bigger tube with the help of plastic struts. One end of the out side cylinder is sealed by an aluminium plate. Through a hole in the centre of the plate is fixed a propeller attached to an electric motor (0.25 H.P.) with a voltage regulator (Dimmerstat) and a reduction gear (a pulley device by which the speed of the propeller can either be reduced to half or increased to double the actual speed of the motor). The propeller blade is just inside the open end of the inner cylinder. An aluminium door provided in the other end of the outer cylinder could be opened to introduce a fish into the inner cylinder of the respirometer. Two stand pipes with slide tubes, at each end of the outer cylinder serve as water inlet and outlet respectively. A third pipe fixed near the outlet tube is used for water sampling. The speed of the propeller is regulated by controlling the voltage supply to the motor. The rotating propeller maintains a current of water through the inner cylinder and back through the outer cylinder to the propeller end. The rate of rotation of the propeller shaft is calibrated against the current velocity in the inner cylinder by a pivot tube. During the experiment, the fish is confined inside the swimming chamber by a set of plastic baffles in front of the propeller and a grid of horizontal plastic bars attached to the door at the back. Two of the bars carry charged electric wires to the grid which is used to prevent the fish from falling back and resting against the grid. When the fish falls back and touches the wire grid, it closes the circuit thereby resulting in an electric shock (3 volts) which effectively stimulates the fish to resume swimming.

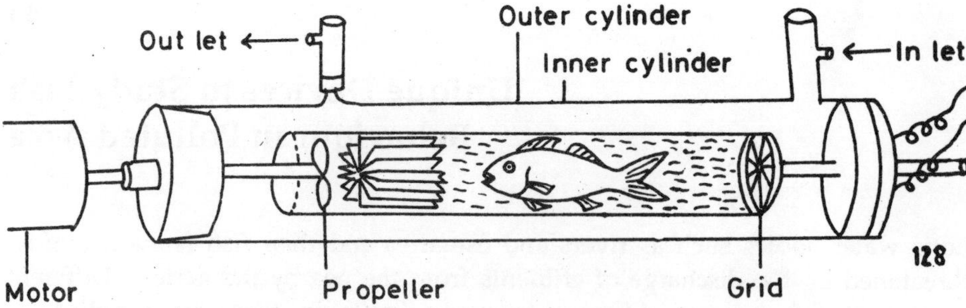

128

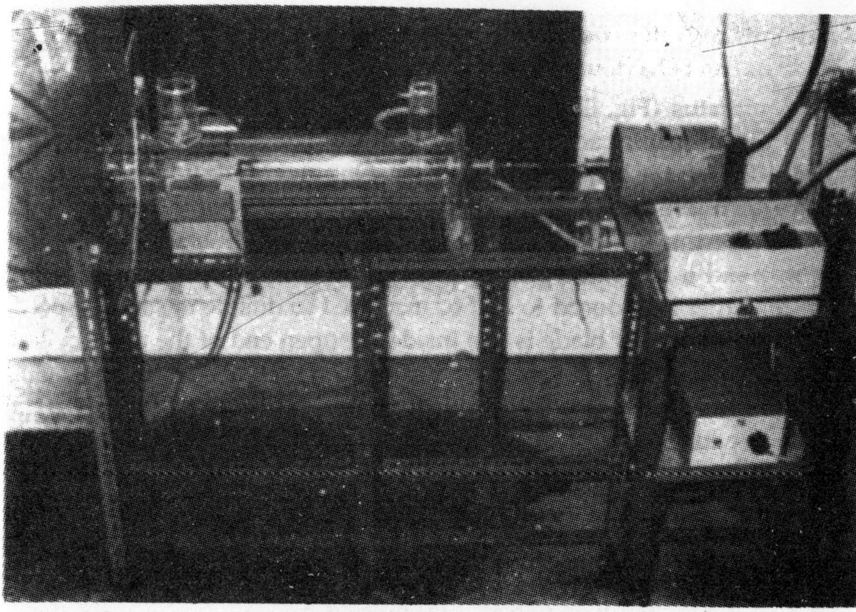

129

Fry Respirometer (Fig. 130) : This is an annular chamber made of transparent acrylic plastic with a photocell system for counting the activity of a fish. The photocells are fixed in the inner hollow of the annular chamber and beams of light are focussed from outside into the photocells. When the fish moves, it cuts the light beam and the event is registered electronically on a chart recorder.

The fish activity counter has also been improved with a photocell system having a light source at the centre. The photocell rack moves outside the annular chamber shadowing the fish. The unit could be adjusted to measure correctly the linear movements of fish.

130

For estimating the energy expenditure of an experimental fish under normal conditions, the fish is allowed inside the sealed chamber of the Fry respirometer containing water for which the initial oxygen content is recorded. After a known time, the oxygen content is again recorded and the rate of oxygen consumed by the fish in terms of mg 0_2/hr/kg body weight is then calculated. For estimating energy utilization of fish in active state (swimming), the same procedure is followed using Blazka apparatus. However, for behavioural studies, instead of quality water, media containing different concentrations of toxic chemicals are normally experimented.

44

Electricity in Fishing

Due to the ever increasing need of cheap protein for which fish is the main source, methods for capture of fisheries on traditional basis will not be of much use in the near future. Unless we find ways and means to modernise the crafts and gears, we may not be in a position to exploit the needy resources within available time.

Among gears (nets) two types viz. active and passive fishing gears are now existing. The former consists of all towed and dragged fishing gears including trawls which have to be moved or manipulated in order to follow and to catch the fish. Passive fishing gears on the other hand are those in which the fish voluntarily surrender. To these, seines, fences, longlines and traps belong. Electricity plays a vital role in modern fishing gears in which the direct current or pulsating current works as an 'attracting and concentrating bait'. Fishing with electric devices could catch more fish and solve energy crisis.

Fish Behaviour in Electric Field

Reactions of fish are almost same in both freshwater and sea water. However, the body voltage is greater in seawater than in freshwater. This is due to the fact that the conductivity of fish body is substantially smaller than that of surrounding seawater. The conductivity of seawater is on the average 500 times greater than freshwater and in the case of seawater, it amounts to $3.10^{-6} A/cm^2$. It is necessary to use therefore 500 times greater voltage in case of direct current to produce similar electric field as in freshwater.

The reaction of fish in electric field depends on the type and size of fish, conductivity of the medium and type and capacity of electricity used. When direct current is passed into a fish tank, all fish tend to swim towards the anode (Fig. 131). But before reaching it, they are usually bewildered and turn upside down. When the current is switched off, they recover and swim away. On the other hand, in the alternating current, the fish do not swim towards one of the electrodes but they take a transverse position to the direction of current between two electrodes.

The effect of the passage of current on the composition of blood of fish is almost nil. For example the RBC, haemoglobin level, erythrocyte, sedimentation reaction, leukocyte formula and glycemia level close to normal. There is no difference in length and mean weight increase between experimental fish (that is fish exposed to electric current) and controls over even half a year. Likewise, there is no difference between experimental and control eggs and larvae of fishes.

Courtesy : Science Reporter

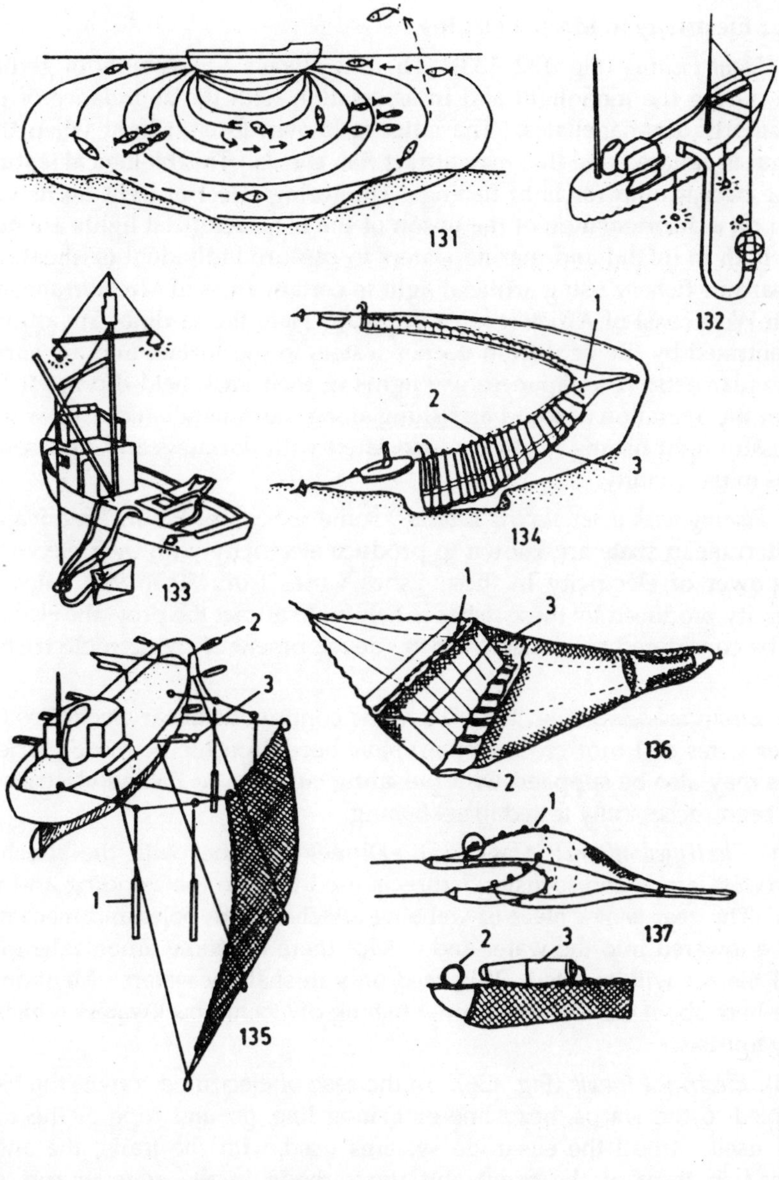

General Uses of Electricity in Fishing

The electric fishing instruments of freshwater are operated by either direct current or alternating current through batteries and that of marine systems mainly by pulsating current through diesel powered generators. The general uses of electricity in fishing are: (i) to increase density of fish shoals; (ii) to improve the retention of fish in certain zones of the gear during discharging the fish outside; (iii) to control the migration routes of fish, and (iv) to help selective fishing for various purposes.

Use of Electricity in Marine Fishing

1. *Light Fishing* (Fig. 132, 133). There is always a possibility of getting good catch during the moonlight and in association with the abundance of plankton particularly dinoflagellates. The latter is known to emit light when the water column is agitated and this may attract fish shoals. This biological feature could be the background for light fishing. The luring effect of light on fish is partly based on a disorientation of the vision of the fish. Artificial lights are nowadays used both in inland and marine waters to capture individual or shoaling fishes. The sardine fishery using artificial light in certain areas of Mediterranean and off North West coast of Africa is most popular. Here the sardines are attracted and concentrated by the light from deeper waters to the surface and are surrounded by circular nets. The Japanese use lights in their stick held dip net fishery and also in the operation of squid jig fishing along the Australian coastal waters. The successful light fishing is always associated with darkness and in absence other lamps in the vicinity.

2. *Fishing with other electric devices.* Some fishes like South American eel and Mediterranean skate are known to produce electricity with their electric organs. The power of electricity in these fishes varies from 50 to 900 volts. As the electricity produced by these fishes is helpful to attract the prey, the electric fishes may be considered as pioneers for the development of modern electrical fishing devices.

(i) *Electrified seines* (Fig. 134). Here the continuous linear electrodes (made of copper wires of 1 mm^2 cross-section) have been found to be more efficient. The seines may also be supplied with pulsating current. In Poland electrified seines have been successfully tested in eel fishing.

(ii) *Electrified lift net* (Fig. 135). Dipnet or liftnet with the attachment of electrodes, lamps and focusing lamps is used both in the standing and mooring boats. The gear is of a piece of webbing attached with poles and mechanically it will be lowered into the water and is kept there for some time. When fish, etc., crawl the net will be lifted. It is used only in shallow waters. Mention may be made here about the Chinese dipnet fishing of Cochin backwaters which attracts many tourists.

(iii) *Electrified trawls* (Fig. 136). In the case of electrified trawls, the electrodes are fixed to the warps, head line or fishing line, ground rope or the netting of trawl itself. In all the electrode systems used with the trawl, the anodes are situated in front of the trawl and the cathode in the zone of cod end. In electrified trawls it is advantageous to use strong cables as warps which are able to withstand the mechanical load and act as conductor. Pulsating current is used to attract fish to the zone ahead of trawl mouth and alternating current is used to drive fish to the line axis of trawling. Both ordinary and pair trawls are electrified nowadays. The radii of capturing action of electric fields of currents are relatively small, therefore the main purpose of electrified trawls is of disorientation of fish, reduction in mobility and inducing fish to fall ahead of mouth of trawl, all of which help more efficient fishing. In the case of bottom and near bottom trawling, the effects of the electric field are important as it drives the

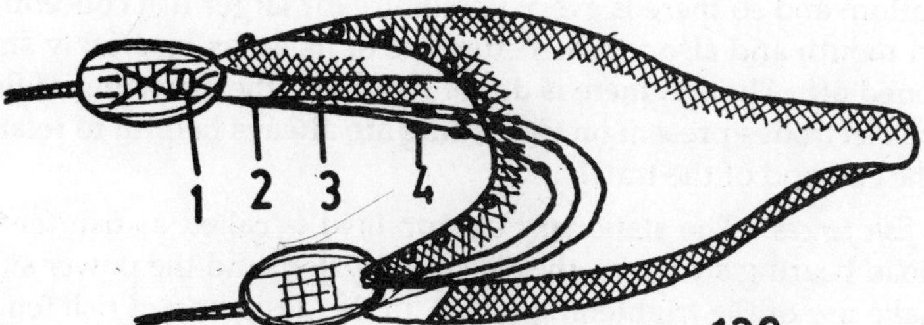

138

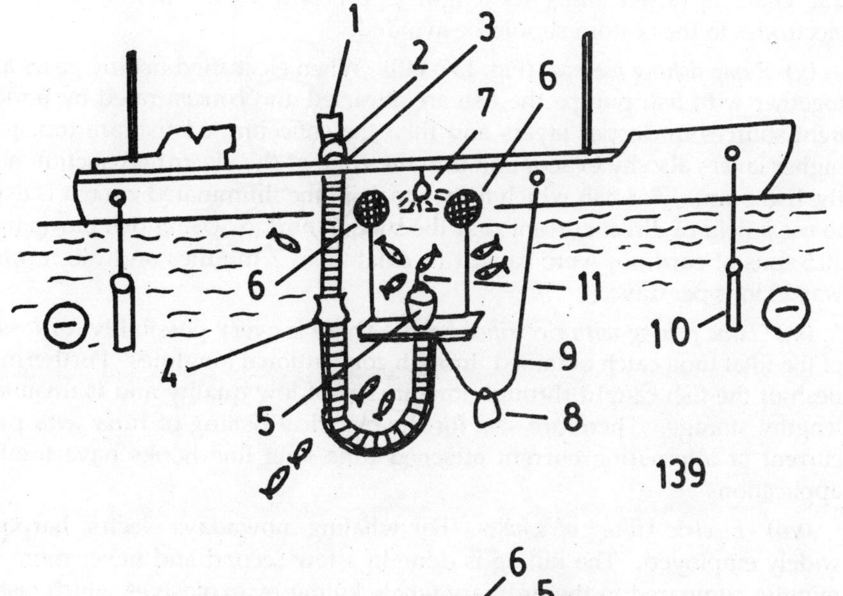

139

140

fish from the bottom and so there is every possibility for larger fish concentration in front of trawl mouth and also for the better size of fishes as invariably smaller ones are frightened off. Though there is doubt regarding the attracting of fish by electric field, the electrodes present on the netting are always helpful to retain the caught fish in the cod end of the trawl.

(iv) *Electric fish fences.* The stationary electric field is called as fish fence. It consists of the load bearing structure, the electrode system and the power supply. These fences make use of the frightening effect. In the designing of fish fences, it becomes necessary to change the distances between the electrodes or between their active surfaces to ensure optimum field configuration and uniform current distribution between the electrodes. Usually metal screens or pipes are used as electrodes. When the water is deep, each electrode is a chain with many links. The chain is raised along its length to conform, to the bottom relief. Fixing electrodes to the bottom should be avoided.

(v) *Pump fishing method* (Fig. 137-140). When electrified fishing gears are used together with fish pumps the fish are attracted and concentrated by underwater light sources in deeper layers and then the concentrated fish are transferred to higher layers also by evacuation into the zone of the electrified suction nozzle of the fish pump. For fish which form shoals in the illuminated zone it is expedient to use a field of direct current near the suction nozzle. Using fish pump methods, 10.5 tons of sardines were caught in Atlantic in 7 minutes and the record catch was 20 tons per day.

(vi) *Tuna fishing with electrified hooks.* There is every possibility for loss of half of the total tuna catch obtained through conventional longlines. Furthermore, the flesh of the fish caught through longlines is of low quality and is unsuitable for lengthy storage. Therefore sets for the electrical fishing of tuna with pulsating current or alternating current attached tuna long line hooks have found wide application.

(vii) *Electric killing of whales.* For whaling, nowadays electric harpoons are widely employed. The killing is done in a few second and never more than 10 minutes compared to the ordinary whale killing by explosives which lasts seven hours. The flesh obtained through ordinary killing may be of low quality. Much lighter lines may be used in electric whaling because the resistance of whales is minimal. The harpoons are fired from ordinary harpoon guns and one pole is fixed to them. The second electrode is in the ship's hull. Whaling with pulsating current is more reliable as the harpoon does not penetrate deeply into the body and it requires less electric power. The harpoon lines are sometimes special coaxial cables whose central core is connected to the harpoon.

Use of Electrical Fishing in Inland Waterbodies

In inland water bodies, electrical fishing helps mainly in aquaculture operations such as: (i) to eradicate undesired fish by electric shock; (ii) to catch spawners for the purpose of induced breeding: (iii) to catch fish like eel which are difficult to catch by conventional methods; (iv) sanitary removal of sick fish; (v) to help assessing fish stocks to get biological information like tagging, etc.; (vi) to help commercial fishing in shallow snagged sections of water bodies where fishing with conventional gears is impossible; (vii) to protect stocked juveniles against predators in various river sections with the help of electric fish fences (viii) to regulate the migration paths of fish by establishing fish fences; (ix) to fish the deeper sections of storage lakes by employing electrified trawls and drag seines; and (x) electronarcosis of fish for transporting alive and for tagging.

Although foreigners have been utilising electrified gears for a number of years, we have not fully used electricity for fishing due to the lack of modernised fishing vessels and gears. However, based on recent research in electrified gears, it is certain that electricity could play a vital role in the fishing gear technology of India for better exploitation in the near future.

45

An Electronic Device for Detecting Fishes

In ancient days, depth of water column in the sea was measured by dropping a line with a weight at its end into water. This method however, is inaccurate as the water is always moving. In modern days, equipment like 'Echo Sounder' play a major role not only in finding out sea depth but also in detecting shoals of fish.

Sound travels in all directions and takes a certain amount of time from one point to another. The velocity of sound is constant for all frequencies in any one medium through which it travels. In air, the velocity of sound is 3.32 m/sec and in water about 1500 m/sec. If the number of vibrations per second of sound (i.e. frequency) is more than 20,000 counts/sec, it is said to be above audible range. Such sound waves are called ultrasonic waves. Sound waves are reflected when they reach a different medium. The reflection actually takes place at the interfaces of two media. For example, in sea, water is one medium and the sea bottom or moving fish shoals is another. As a result, part of the reflected waves return as echoes to the point of the origin of waves. In actual practice, an ultrasonic wave of very short duration, of the order of a small fraction of a second, is transmitted from a place and the reflected echo is identified. The time interval between vertical transmission of the sound and reception of its echo is a measure of the distance to the reflecting surface. This is the basic principle on which an echo sounder works.

A typical echo sounder consists of two parts, viz., cabinet and the transducer. The cabinet, located in the wheel house, consists of the recorder, the receiver and the transmitter (Fig. 141). Recorder is the brain of echo sounder because it starts the transmitter and records echo after receiver has amplified approximately 1,000,000 times. The transducer is fitted under the hull of the ship and acts as a loudspeaker for the transmitter and as a microphone for the receiver. The echo sounder has been designed to work at supply voltage of 12 V or 24V battery using lead-acid accumulation.

In the recorder unit echoes are recorded by means of a pen passing over a special recording paper. When the recorder directs the transmitter to send out a sound beam through the transducer, the pen in the recording unit makes a 'zeromark' on the paper. The mark indicates the moment sound beam leaves the transducer. As the sound beam travels down through water, it is reflected by organisms such as fishes and plankton below the boat and the bottom of the ocean. Some of these reflected echoes are gathered by the transducer which now acts as a microphone. The echoes from the objects near the boat are recorded as echo mark closer to the zero line. Echoes from objects farther away take a longer time to come back so they do not arrive until the pen has already been on its way

Courtesy : Science Reporter

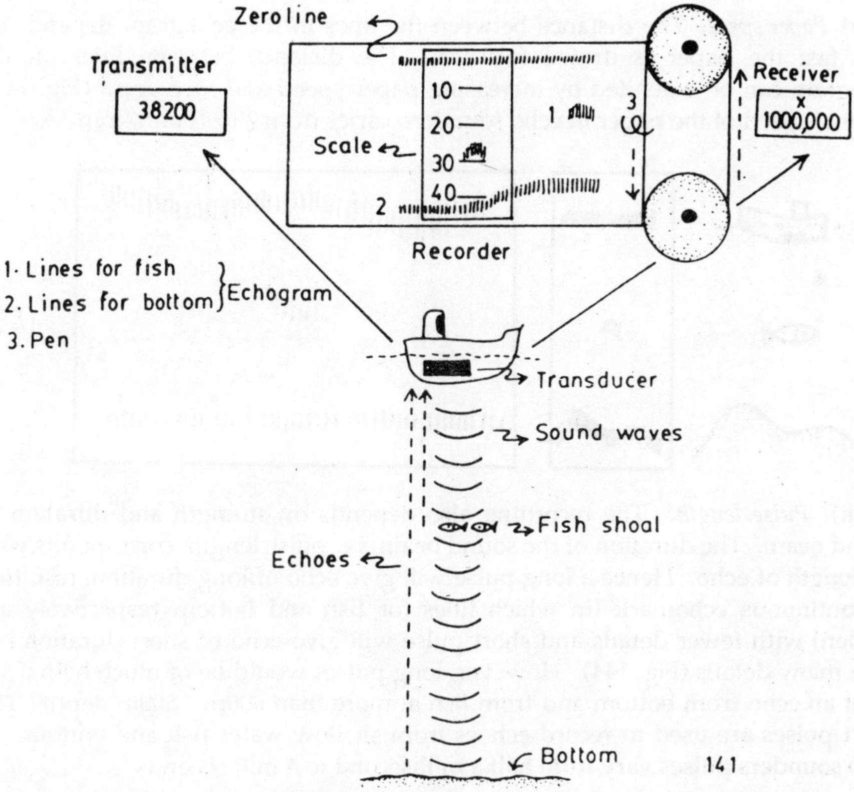

141

across the paper. By setting a scale over the paper one could read the depth of the different objects and of the sea bottom. The scale is based on the speed of sound waves in water. At the depth of 1500 m, the sound pulse requires one second to travel from boat to sea bottom and requires another second to travel as an echo buck from it. The echogram however depends on various other factors such as (i) pen speed, (ii) recording paper speed, (iii) strength and duration of sound beams transmitted, i.e. 'pulse length', and (iv) phasing.

(i) *Pen speed.* If the speed is high, the pen travels farther in the short space of time and makes a longer mark in the echogram and vice versa. (Fig. 142).

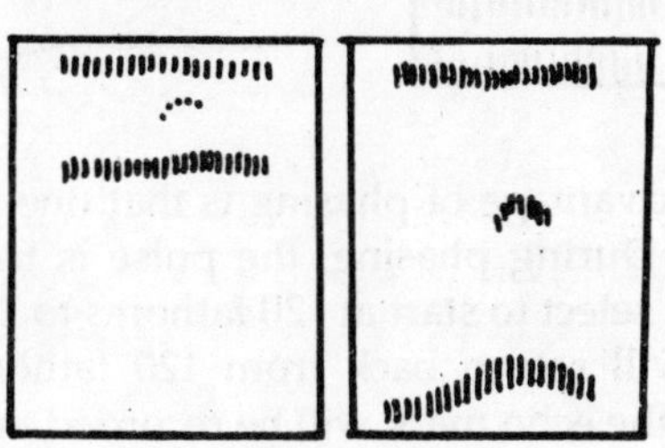

142

(ii) *Paper speed*. The distance between the lines in the echogram depends on how fast the paper is drawn forward. The distance between lines on the echogram can be extended by increasing paper speed and vice versa (Fig. 143). Normal speed of the paper in echo sounders varies from 2 to 10mm/min.

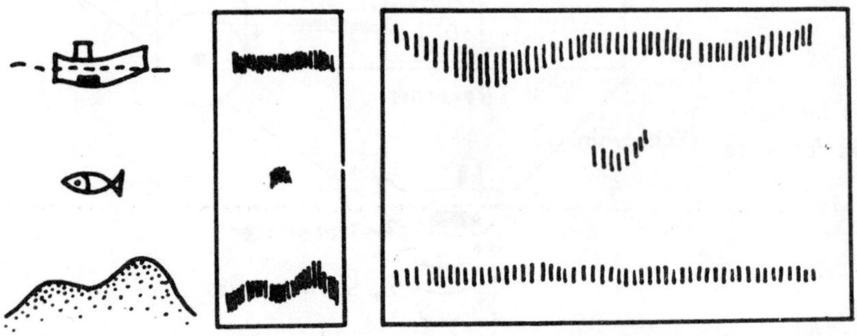

143

(iii) *Pulse length*. The recording also depends on strength and duration of sound beam. The duration of the sound beam, i.e. pulse length, corresponds with the length of echo. Hence a long pulse will give echo of long duration, resulting in continuous echomark (in which lines for fish and bottom respectively are hidden) with fewer details and short pulse will give echo of short duration but with many details (Fig. 144). However, long pulses would be of much help if we want an echo from bottom and from fish at more than 600m—800m depth. The short pulses are used to record echoes from shallow water fish and bottom. In echo sounders pulses vary from half a millisecond to 4 milliseconds.

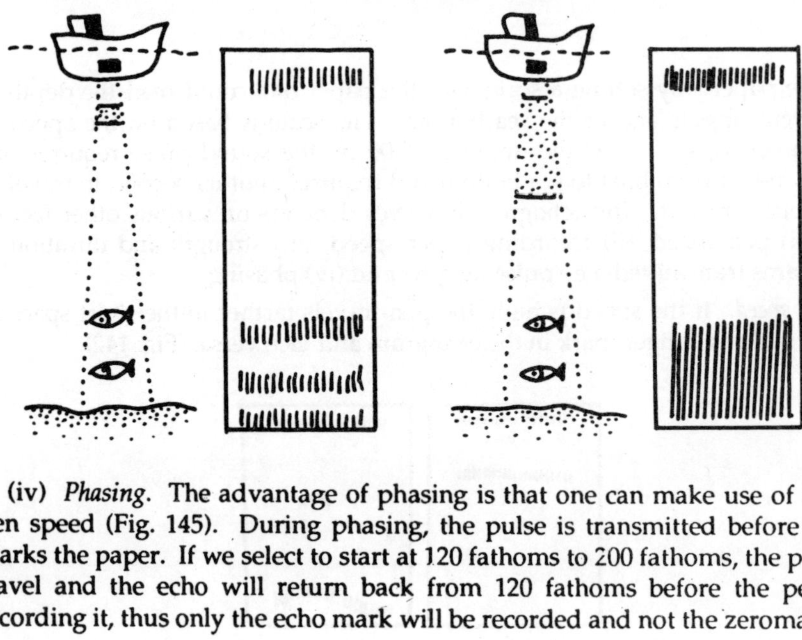

144

(iv) *Phasing*. The advantage of phasing is that one can make use of the high pen speed (Fig. 145). During phasing, the pulse is transmitted before the pen marks the paper. If we select to start at 120 fathoms to 200 fathoms, the pulse will travel and the echo will return back from 120 fathoms before the pen starts recording it, thus only the echo mark will be recorded and not the zeromark.

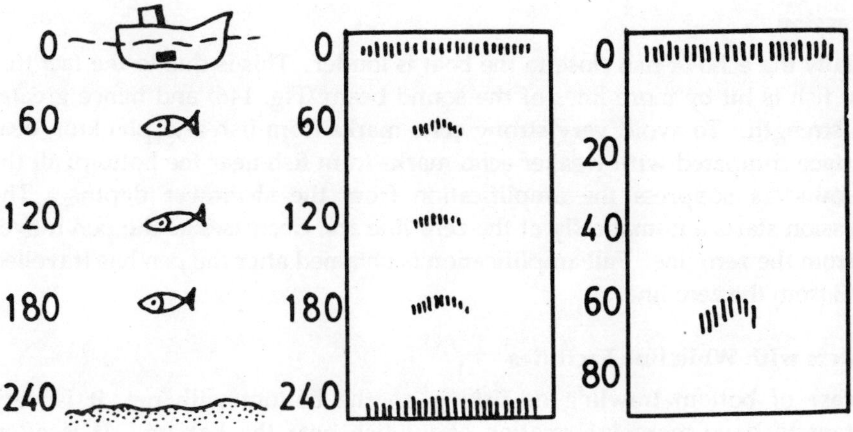

145

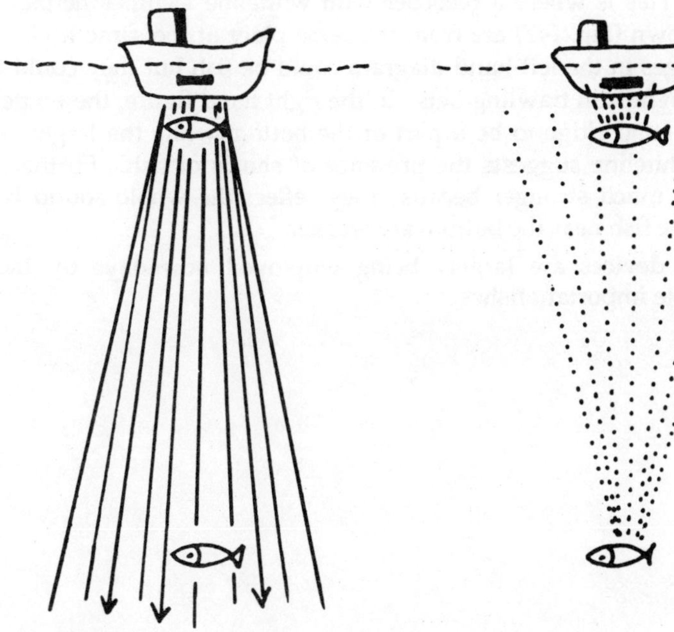

146

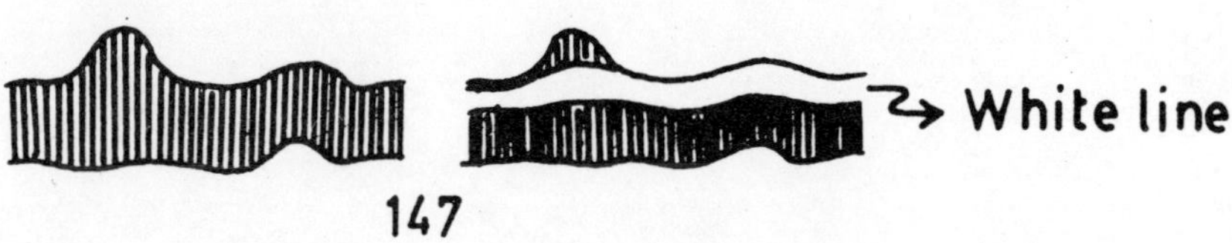

147

Suppression

Usually the echo of fish close to the boat is louder. This is due to the fact that nearest fish is hit by more lines of the sound beam (Fig. 146) and hence greater sound strength. To avoid very strong echo marks from fish and plankton near the surface compared with weaker echo marks from fish near the bottom, all the echo sounders suppress the amplification from the shallower depths. The suppression starts automatically at the zero line and decreases as the pen moves away from the zero line. Full amplification is obtained after the pen has travelled 1 to 1/3 from the zero line.

Recorders with Whiteline Facilities

In case of bottom trawling or fishing on the bottom with net, it is very important to have more information about fish near the bottom. It is often difficult to distinguish echolines of fishes near bottom than from the bottom itself in the echogram. This is where a recorder with whiteline facilities helps. The two echograms shown (Fig. 147) are from the same place at the same time. The two hump-like ridges in the left hand diagram could be fish but they could also be stones which might ruin trawling nets. In the right hand figure, the whiteline suggests the right hand ridge to be a part of the bottom while the larger ridge lying above the whiteline suggests the presence of shoals of fish. Further, the bottom echoes are much stronger because they reflect the whole sound beam while those from the fish near the bottom are weaker.

Such electronic devices are largely being employed nowadays by Indian vessels to help locate important fishes.

46

Devices to Lure Fish for Bumper Catch

Fish aggregating devices (FADs) are either natural or man-made. They are suspended in aquatic bodies such as sea, estuaries, backwater, mangrove waterways and reservoirs to help attract and concentrate pelagic, demersal and migratory fish such as sardines, mackerels, tunas, carangids, sharks, cat fishes and crustaceans. FADs are considered modern, efficient and fuel saving.

A typical FAD has three components (Fig. 148) viz. (i) raft (ii) attractants and (iii) mooring line and anchor. It consists of a long coir rope buoyed at the upper end with a float of light wood and moored with a 25 kg stone. The first 20m from the bottom is kept free. Above this height, 50 coconut or palmyra branches are strung at intervals of 50 cm. After two weeks of mooring, attracted fish shoals are caught with a scoopnet operated by fishermen from four catamarans. Sturdy FADs on the other hand consist of rafts, made of (i) bamboo clusters held together with chain links, wires or used car tyres (ii) aluminium hulls (iii) Polystyrene foam-filled (inside) truck tyres and (iv) high density polyethylene (HDPE) tubes of 7.5-15 cm diameter with hermetically sealed ends.

A raft may also be made of three 200 litre empty oil drums welded in a frame of angle iron. The raft is also provided with a short mast, blinking light and flag for easy location. Automobile tyres or tufts of black or green coloured plastic strips tied in a vertical fashion in a hanging polypropylene rope (20mm) from the raft serve as attractants. Alternatively, nylon nets or old trawl net pieces may be used. Both the raft and attractants are moored by polypropylene rope—galvanised iron chain (30-50 mm) combination and anchors made of concrete-filled oil drums. For a large FAD, an anchor weighing 1,500 kg may be needed. Fish aggregated in such devices are caught using handlines, long lines or surrounding nets such as entangling nets and seine nets.

Bottom-living fish, crustaceans and octopi may also be attracted by three dimensional structures which provide not only food and refuge but also serve as breeding grounds.

Artificial parks: Dense clusters of twigs or branches of trees are planted vertically using stakes in the muddy bottom of estuaries and mangrove water ways influenced by tidal waters. These artificial parks are known to attract fish.

At intervals, these heaps are shaken or slightly lifted after the site has been encircled by gill nets or traps to catch the fish when they endeavour to escape. The entire structure may also be completely removed and the trapped fish caught by different methods. Regular feeding will familiarise the fish with the place.

Courtesy : The Hindu

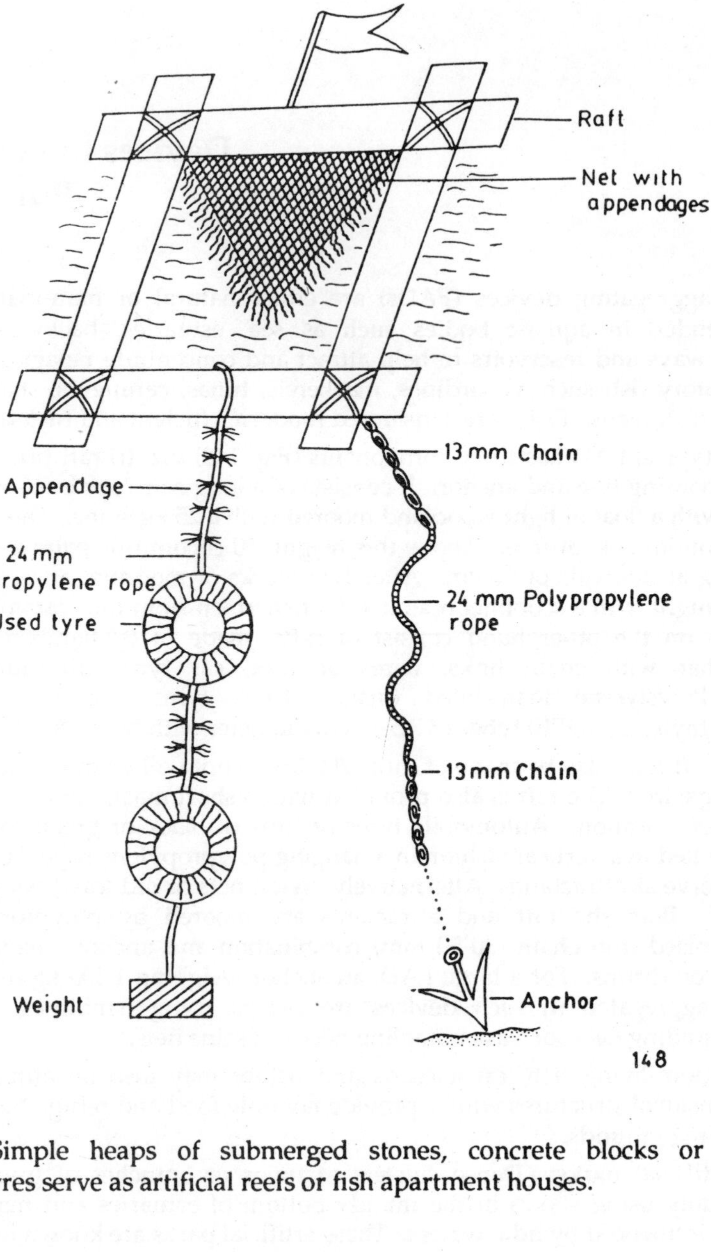

148

Reefs: Simple heaps of submerged stones, concrete blocks or used automobile tyres serve as artificial reefs or fish apartment houses.

47

Solar Driers for Hygienic Drying of Fish and Farm Produce

The use of solar driers has been considered to be a suitable alternative now-a-days to traditional sundrying of fin fishes and agricultural produce in tropical developing countries. These driers are capable of collecting and concentrating solar radiation to achieve elevated temperatures and reduced relative humidities during drying. This results in increased drying rates, lower final moisture content and higher quality products. Further, it is less susceptible to variations in weather, apart from providing shelter from rain. Pests and predators do not enter the driers due to the high internal temperatures which are lethal to such animals.

Although many designs of solar driers are available for the use of agricultural produce, only a few are employed for drying of fish and fishery products.

Description of a Moderate Sized Solar Tent Drier (Fig.149)

It consists of a bamboo frame of suitable size. While black polyvinyl chloride (pvc) sheet is used at the base and ends of the drier, clear ultraviolet resistant

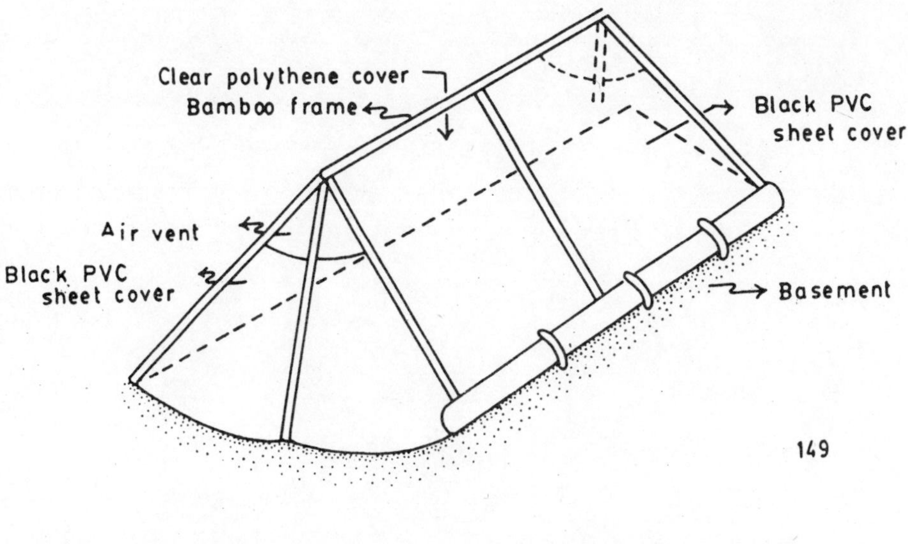

149

Courtesy : Farm Digest

polyethylene sheet is used for the top sides. A movable plastic flap forming half of one end of the tent serves as mouth of the drier for introducing materials for drying or for the removal of dried materials. Solar radiation which passes through the clear plastic sides of the tent is absorbed in the black plastic sheet thereby increasing the temperature of the inner areas of the tent. Further, the air which enters through the openings at the sides of the base is heated and rises, thus inducing a drought within the tent. Vents in the apex at both ends allow the used air to exhaust. Internal temperature and air flow through the drier, can however be controlled by adjusting the height of the side openings.

Advantages of Solar Driers

Fish dried in solar driers are known to take an average of about 60-65% of the time required for traditional sundrying to achieve a final moisture content of 20%. Further, under normal weather conditions, drying could be effected in a period of three days in the solar driers compared to five days in traditional sundrying. It is also possible to achieve a greater reduction in moisture content by solar drying (up to 13% on average) than by ordinary sundrying (21% on average). Products obtained hygienically in solar driers have also been found to hold a much longer shelf life.

48

Prawn Picking—A Sustenance for Rural Women Folk

India, apart from its seas and freshwater bodies, also abounds in many shallow brackishwater environments such as backwaters, estuaries and mangroves, the extent of which is about 1.5 million ha. These systems are known for their rich fin and shell fish resources as these less saline and organic-detritus-rich systems serve as feeding and spawning grounds for many species. However, their exploitation in most of the localities is mainly by sophisticated nets.

'Handpicking', otherwise called 'fishing by foot and barehands' is an age-old practice and it has led to the development of all other modern fishing techniques. It is in vogue in many shallow coastal localities by fishermen especially for the collection of molluscan shellfish for lime production. But handpicking of prawns using simple indigenous devices by wading women is unique and it is practised only in limited areas like Pulicat lake and Pullavazhi (near Tuticorin) in Tamilnadu.

As prawns, *Penaeus indicus, P.semisulcatus P. monodon* and *Metapenaeus dobsonii*, migrate to the brackishwaters and live submerged in the muddy bottoms, catching these sluggish and delicious animals by handpicking is a fascinating art that stands apart from any other harvesting methods.

Handpicking of Prawns in Pullavazhi

Rural women in the age group of 15-75 and numbering about 50 regularly engage themselves in this avocation. Besides, a number of salt pan workers within a radius of about 20 km also join in this venture as an alternate job particularly during rainy season. Women enter into these waters during low tides and the operation lasts for about 5 hrs, 9 am to 2 pm daily. Before the start of the operation, all the women form a queue between the two banks of the area, move very slowly and grope carefully along the muddy bottoms, keeping their heads above water column. Each one of them has a mud pot containing twigs which is tied to their hip with a small rope. The pots float in front of them (in Pulicat lake, women carry palmyra leaf bags on their backs). When a prawn is found fit it is held firm by the feet till it can be grasped with the hands. The hand-picked prawns are put then and there on the floating pots and the twigs of the pot help prevent the escape of the prawns from the pot. As soon as the women reach the shore, the catches are sold to the waiting fish mongers who fix the price according to the size of the prawns landed. Every woman gets a daily income Rs. 8-15 depending on the seasonal conditions.

Courtesy : Fishing Chimes

Besides providing a livelihood, this shore based operation is of importance as it helps to strengthen social relations between the women handpickers of each locality. In Tamilnadu, where brackishwater areas are also available in Muthupet, Covelong, Marakkanam, Killai, Ennore, Adyar, Pudupattinam, Cuddalore, Parangipettai, Coleroon, Nagore, Vailankanni and Nagapattinam, the handpicking of prawns, if popularised among the rural women folk, will certainly serve as a sustenance fishery without any major investment.

49

Fishery Byproducts of Commerce

India's marine fish production has now crossed one million tonne mark. Of this, trash fishes constitute more than 40 per cent. These are either thrown back into the sea immediately after their catch or utilised wholly or partly as manure. If this fraction is used in the preparation of many usable by-products, it would certainly augment our foreign exchange earnings.

Chitosan

The left-outs of prawns prior to their freezing in the processing factories such as their head and shell-wastes are now used mainly as manure or as a source material in the preparation of feeds. From these wastes, Chitosan, a valuable by-product can be extracted. In its preparation, three steps are followed, namely, deproteinisation, demineralisation and deacetylisation. In the deproteinisation, the prawn wastes are first washed and heated with 3% sodium hydroxide solution for one hour in order to remove the crude protein content of these wastes. Subsequently, it is washed well and transferred to a vessel containing 5% hydrochloric acid which removes the mineral content of the wastes in a period of about 30 minutes. This process is known as demineralisation. The product so obtained is Chitin which contains 60% moisture. The Chitin is then subjected to deacetylisation. In this process, the Chitin is washed well and heated with 40% sodium hydroxide solution for about one and half hours. Then the material is washed well, sun-dried and pulverised. The resulting chitosan should be packed in polyethylene bags and stored at ambient temperatures. Chitosan is nowadays used as a valuable Coagulant aid in the treatment of sewage, etc. It is also used as an additive in the baby food formulations; treatment of wounds; paper and textile industries and in photography.

Shark Fins

The translucent cartilaginous rods embedded in the fins of shark are the fin rays used in the preparation of shark fin soup. These rays can be extracted from both freshly cut as well as dried fins. The latter are soaked in water which is acidified with acetic acid with pH 2.5 to 5 for 2-3 days while freshly cut fins require less soaking time. The softened fins are then treated with hot 10% acetic acid at 60°C for an hour depending upon their size. The rays are separated manually, washed well and draied in the sun. The dried rays which can have a moisture level of 5-8% are stored in polyethylene bags. The shark fin soup is considered as a delicacy in countries like China, Philippines, Hongkong, Singapore, etc.

Courtesy : Fishing Chimes

Beche-de-Mer

Holothurians, popularly known as sea-cucumbers, form the raw material for the production of the commercially important Beche-de-mer. The industry, however, at present, is mainly based on a single species viz. *Holothuria scabra.* The animals belonging to this species are first washed and heated in a vessel without water in order to remove their internal organs. Then they are washed and again heated for about one hour till they get shrunk to half of their original size. The materials are then buried in sand and kept for 18 hours. Afterwards, the de-skinning is done and again heated with seawater for 15 minutes. Subsequently, they are dried well in the sun. The product is dark-red in colour and is exported to Far Eastern countries like Singapore, Hong Kong, Malaysia, etc.

Fish Oils

Fish oils are of two types viz. fish body oil and fish liver oil. In India fish body oil is mainly prepared from the oil sardines belonging to the species **Sardinella longiceps** and liver oil from shark, skate, tuna, ray, etc. Shark liver forms a very rich source of Vitamin A and D. In India, at present the products are prepared at Kozhikode, Trivandrum and Bombay centres. The efficiency of extraction is more in the alkali digestion method. The fresh or preserved livers are washed and weighed. Then the associated structures of livers such as gall bladder and veins are removed and washed again. Subsequently livers are cut into pieces to which sufficient quantity of water is added and the mixture is digested at 40-45°C till all the solids of the liver are liquified. The digestion is also continued with 1-2% sodium hydroxide at 40 to 90°C. The pH is adjusted to 8 to 9 with H_2SO_4 or HCL solution with constant stirring. To the cooled emulsion 3 volumes of 5% saline solution are added. Afterwards, it is centrifuged and washed in warm water. The separated oil is mixed with about 5 to 10% anhydrous sodium sulphate and is left undisturbed overnight. Finally, the product is filtered and stored in suitable airtight containers. Fish body oils are used in the manufacture of detergents, rubber, lubricants, printing inks, leather and cosmetics.

Fish Meal

Fish meal is a nutritious feed supplement consisting mainly of proteins, minerals and vitamins. Trash fishes such as leiognathids, lesser sardines and anchovies and fish wastes are commonly used for the production of fish meal through wet reduction process. In this process, the aforesaid source materials are ground and cooked. This would facilitate coagulation of protein and release of water and oil. Afterwards pressing is done to separate water and oil. The liquid portion is known as press liquor which is passed through a screen to remove solid particles of fish. Afterwards these particles are returned to the press cake. After drying the press cake, it is ground well and packed in polyethylene-lined gunny bags.

Constituents of the meal vary depending on the type of raw material and the process used. Protein is generally around 65%. Moisture, fat and ash content vary at 6-10%, 5-10% and 12-33% respectively. Fish meal is also known to contain good quantities of all essential and non-essential aminoacids. Fish meal is also an important source for minerals such as calcium, phosphorus, copper, zinc, manganese, iodine and selenium. Hence fish meal is considered to be a valuable constituent of fish and poultry feeds.

Fish Protein Concentrate (FPC)

It is a colourless and odourless powder with highly concentrated animal protein and minerals. It is also known as fish flour, or edible fish meal. In its preparation, the minced meat is cooked with an equal volume of 0.5% acetic acid and is allowed to settle. The oil which floats on the surface is removed. The slurry is filtered through canvas bags and is pressed. The pressed cake is treated first with ethanol to remove both moisture and odour and then with azeotropic mixture of hexane and ethanol. The treated mixture is subsequently filtered, pressed and dried under vaccum. Finally it is pulverised and packed.

As it is a stable compound for a period of even 3 years at room temperatures, and is rich in protein it can be incorporated in several food preparations like chapathi, poori, bread, biscuit, etc at 5 to 10%.

Fish Glue

Thick and firm skins of certain fishes are generally suitable for the preparation of fish glue owing to their good collagen content. However, low fat content of skins is desirable for the preparation of quality fish glue. The skin is washed and soaked in freshwater to remove salt or dirty substances. After soaking, they are cut into pieces and transferred to lime liquor which helps to open the fibre bundles. After washing, the limed skin pieces are bated to remove lime. Afterwards, it is cooked in steam cooker and a small amount of acetic acid which acts as a catalytic agent is added to hasten the hydrolysis. Then the glue liquor is drawn and evaporated until the liquid contains 50-55% solids. Afterwards, it is cooled below 15°C in a cooling box for 12 to 18 hours to coagulate. The coagulated jelly type glue is cut into moderate sizes and then dried in a drier. Any one of the antiseptics such as boric acid, Salicylic acid or benzoic acid is added to liquid glue in order to conserve the product from spoilage. It is used for fixing wood, paper and leather. It is also used in book binding and labelling.

Fish Pulp

Sardine, mackerel, sciaenids, etc., are used in the preparation of fish pulp. The meat is separated first from the skin and bones of selected and leaned fish and is treated with powdered salt (33% by weight). The brine that appears in a duration of about 6 hours in the above mixture is removed by pressing the meat, thereby the moisture content of the fish is reduced to 40%. Removal of the brine and the associated pigments help to increase the quality and shelf life of the product. The minced fish meat is then covered again with salt and kept in a hermatically sealed plastic container and stored at room temperature. The high salt content (24%) and intermediate moisture level of the product (50%) inhibit the development of micro-organisms, thereby the shelf-life of the pulp is retained for six months without any preservative. The fish pulp is known to be a good and protein rich food source for the people of many developed and developing countries although it has not yet become popular in India.

Fish Paste or Dough

In the preparation of fish paste or dough, elastic minced fish meat is desirable. The other source materials in the preparation of the product include potato or wheat, starch, salt, sugar, monosodium glutamate and chopped vegetables like carrot. After removing the head, scales, and viscera of fish, the flesh is cut into single fillets and bones and skin are removed manually. The meat is washed three to four times thoroughly in freshwater to remove the fat. Excess water is removed and meat is kneaded in a machine for about 15 minutes. The meat is again kneaded twice for 15 minutes by mixing first with 20-40 g of common salt per kg of meat and subsequently with potato or wheat starch at 100-250g per kg of meat. In order to improve the flavour of product, sugar may be added (at the. rate of 30-100 g/kg of meat) and the kneading may be continued until a dough consistency is obtained. The resulting one is a stiff paste. Sometimes, the resulting paste is broiled on a heated iron plate. It may also be frozen in a contact plate freezer at a temperature of – 40°C.

Fish Sauce

Fish sauce is obtained from the mixture of fish and salt (3:1) which are fermented in wooden vats or concrete tanks. Generally trash fishes and cannery wastes such as head, viscera, fins, etc., could be used in its preparation. After the fermentation period of about six months, the supernatant liquid i.e. fish sauce is decanted and used for human consumption. The residue, otherwise called as fish paste, is also used for human consumption. The colour of the fish sauce varies from straw yellow to amber depending on the species and period of fermentation and its protein content has been found to be varying from 10 to 15%. It is a simple process and involves low capital for its production. Fish sauce, a preferable food item of the people of South East Asian Countries may be used as a condiment in rice dishes and vegetables.

Isinglass

It is a high grade collagen produced from the air-bladders (Swim bladders) of catfish, carps, eels, polynemids, sciaenids, sea bass, etc. The bladders are first removed from the selected fish and the blood and adhering fat materials are scraped off. They are then cut open and washed thoroughly in running water. Then, the outer black membrane is removed by scraping. Subsequently, the bladders are cut into pieces and are dried in an artificial drier or in sun and stored in suitable containers. Isinglass, so prepared is used for clarifying beverages like wine, beer and vinegar. Isinglass also reduces 2 to 0.05% of the suspended solids in beer and increases filtration rate from 3000 to 11,000 litres.. It can also be used as an adhesive base and in confectionary products, Indian ink and as an efficient adhesive for glass, pottery and leather. Products with less than 8% moisture content are however, preferred for industrial purposes.

Ensilage

It is prepared by converting the entire trash fish or its offal into animal feeds by chemical preservation. The product has an advantage over fish meal as the vitamins are not affected in this product which is also free from fish odour. The fresh raw materials are minced and about 10-15 kg of it is placed in a suitable acid resistant container. The meat is acidified first with 3% formic acid or a mixture of sulphuric acid (50%) and formic acid (85%) in order to maintain a pH of about 3. The container is covered and is left (undisturbed) for 3-4 days. The mixture should also be stirred daily to bring about quick liquefaction. After liquefaction, oil removal may be necessary if fish with high fat content are used.

Since the ensilage is in liquid form, handling and transportation costs are considerably high. Hence the ensilage may be converted into solid feed materials by mixing it with rice bran, tapioca powder, seaweed powder etc., in appropriate proportions and drying. This product is also yet to be popularised in India.

Pearl Essence

Guanin is deposited in the epidermal layer and on the scales of most of the pelagic fishes. Since those crystalline guanin are suspended in a suitable solvent in the manufacture process, the product is called pearl essence. The scales are collected from pelagic fish such as sardines, mackerels, carps, mullets and ribbon fishes and are preserved in 10-15% common salt solution until they are processed. The scales are soaked in gasoline to separate pearl essence from protein and water. The pearl essence is transferred into gasoline where it floats to the surface. The separated pearl essence is then filtered to obtain the fine particles of essence. The precipitate is poured again into gasoline. It is used for photography, ash trays, fishing rods, book covers, textiles, jewellery boxes, umbrella handles and electric light switches.

50

Industrial Uses of Prawn Shell Wastes

India earns foreign exchange to the tune of some Rs. 350 crores annually by exporting processed prawns totalling 75,000 tonnes. At present, however, only a negligible quantity of prawn wastes from the processing factories is used as manure and much of them goes waste. These wastes could be processed and converted into either 'chitosan' or fish meal.

1. Chitosan: Fresh or dried shell and head wastes of prawns are employed in the preparation of chitosan. The wastes are washed thoroughly in water to remove the adhering sand, dirt, etc., followed by draining off the excess water. The different phases in the preparation of chitosan are :

(i) Deproteinisation: The protein content of the shells is removed by heating them with adequate quantity of 3% caustic soda (sodium hydroxide) solution in a mild steel vessel.

(ii) Demineralisation: The residue obtained above is transferred to a plastic container and a sufficient quantity of 5% dilute hydrochloric acid is added. This is then allowed to remain for about 30 minutes with occasional stirring using suitable wooden ladles. The excess acid containing the minerals of shell wastes is then decanted. The residue so obtained is washed with water repeatedly to free it from any adhering acid. The acid-free residue is then collected in a canvas or gunny bag and is pressed under a screw press to remove the water content as far as possible. The product obtained now is 'chitin' which may have a moisture content of 60%.

(iii) Conversion of Chitin into Chitosan: The chitin is transferred to a mild steel vessel containing sufficient quantity of 40% caustic soda solution. The contents are heated at 90°C for about 1, 1/2 hours. The unreacted alkali is then drained out as quickly as possible. The water content of the residue is removed as in phase ii and the resulting product is 'chitosan' which is dried in bright sunlight for 6-8 hours. If dried in an artificial drier, the temperature of the latter may be kept around 60°C. The colourless flakes of chitosan so obtained may be powdered using a pulveriser.

Yield and uses of Chitosan: The yields of chitosan from fresh and dried prawn shell wastes have been estimated as 3% and 12% (by weight) respectively. The chitosan has got a number of industrial uses. It is used as an industrial flocculent and as a sizing material for cotton wool, rayon and other synthetic fibres. It also finds application in the preparation of cosmetics, pharmaceuticals

---

Courtesy : Shipping & Marine Industries Journal

and adhesives. It is also shown that treatment of chitosan under special condition is beneficial in giving a permanent organdie type finish to cotton fabrics and also in preventing shedding of jute. Chitosan is also nowadays used in improving the quality of wrapping paper, paper board, butter paper, etc.

2. Fish Meal: The prawn shell waste contains about 45% protein and 5% fat and serves as a good source material in the preparation of fish feed for aquaculture purposes. The waste may be dried in hot sun and powdered for use as an ingredient. A mixture of prawn shell waste, rice bran and groundnut oilcake at 2:1:1 forms an ideal food source for growing prawns in coastal fish farms.

51

Fish Sauce—A New and Promising Byproduct of Commerce

It is estimated that the trash fishes constitute more than 30% of the total landings and at present much of them is either utilized for the production of fish meal or as organic manure. Now it has been understood that this fraction could be effectively utilized for the production of fish sauce for human consumption with long shell life following easy and low cost technology. Though many countries like Philippines, Indonesia, Thailand, Japan, Malaysia, Kampuchea, Vietnam, Burma and Laos are known to produce fish sauce on commercial line for human use, the importance of fish sauce is hitherto unknown to India.

Species of Fish Employed in Fish Sauce Production

Depending upon the seasons and landings, low priced fishes such as leiognathids, thryssa, sciaenids, carangids, sardines, mackerels, anchovies, ribbon fishes etc., could be used for the production of fish sauce.

Characteristics of Fish Sauce

Fish sauce is basically a fermented product obtained from the slow digestion of a mixture of whole fish or its components such as gills, viscera, head, roe, etc., and salt. It is clear liquid of straw yellow or amber colour. It has a fishy odour, salty taste and cheesy flavour. Depending on the species used, the fish sauce will have varying concentrations of water soluble substances such as peptides and aminoacids.

Methods of Preparation

In the formation of fish sauce through formentation, the enzymes of the fish tissue hydrolyse the protein. The microbes are known to play a minor role, as their density decreased steadily as the fermentation continues. However, the halophilic bacteria do contribute to the flavour of fish sauce.

At home, the fish sauce is produced by mixing uniformly 2 to 3 parts of fish to 1 part of salt (v/v). The mixture is placed in good quality porcelain jars or glass jars with lids and is allowed to undergo very slow fermentation. Occasionally, it is stirred to avoid the microbial contamination of fish. When the fish is almost completely digested, the supernatant liquid, i.e., the fish sauce is filtered and used.

Courtesy : Seafood Export Journal

The large scale preparation of fish sauce is as follows. Uneviscerated fish, viscera, gills or roe is mixed with salt in the ratio varying from 3:1 or 4:1 in wooden cylindrical vats or cement tanks. The containers are provided with one or two taps at the base which help in the draining off the sauce. The inner opening of the taps is filled with cotton to help in drawing clear fluid after filtration. Weights are placed on the lids to keep the fermenting materials below brine. The contents are allowed to ferment for a period of 5-8 months depending on the species of fish and quality of fish sauce. After fermentation, the clear supernatant liquid is skimmed off from the top or drained through the taps of the container. After the first liquid is drawn off, the contents are mixed with salt-water and the fermentation is continued. A total of 3-4 extractions are accomplished. Finally, the residue is taken out of the container and is transferred to a boiler to be finally extracted with boiling brine. The extracts so obtained are mixed with varying quantities of first extract to produce different grades of fish sauce. The filtered liquid is exposed to sun for 2-3 months in jars before bottling. Afterwards, it is packed in glass bottles. From 100 kg. of trash fish about 70 kg. of fish sauce could be obtained. The fish sauce does not require any special storage and can be kept for indefinite period depending on the final salt content of the sauce.

Uses

The fish sauce is widely used as a condiment and as a substitute for solid salt. As it contains considerable quantities of vitamin B_{12} and protein, a daily consumption of 40 ml of fish sauce would substitute about 7.5% of an individual's intake of protein in this could assume a protein intake of about 54 kg per day. In foreign countries, the fish sauce is consumed alongwith boiled rice and vegetables.

52

Ambergris

The occasional appearance of a waxy substance on the surface of the sea and all along the beach in some regions has long been a puzzle. This substance has been termed as 'ambergris' and is sent out specifically by male toothed whales, i.e. sperm whales. Ambergris is said to be formed in the intestine due to the difficult digestion of food particularly the molluscan squids and octopi. The presence of the aforesaid fragmented food in the ambergris confirms this. Formerly ambergris was supposed to have come only from diseased or emaciated sperm whales. But recent research confirms that healthy sperm whales may also produce this substance. The colour of ambergris generally varies from dirty white to blackish brown. A spectrophotmetric examination of ambergris showed the presence of elements like magnesium, calcium, sodium, copper, iron, silicon, zinc, aluminium and nickel. During 1953, mariners of United Kingdom caught a sperm whale and collected 420 kg of ambergris from the intestine. The following constituents were reported by them in the chemical analysis of ambergris.

Ether soluble material	79.8%-97.5%
Ambreine	13.6%-39.8%
Ash	0.7%-8.6%
Epicoprostanol	3.4%-28.0%
P_2O_5	0.1%-2.8%

Though odourless, ambreine is said to be responsible for the fixing properties of ambergris. On the other hand, Epicoprostanol possesses a musky odour.

The use of ambergris is varied. It was previously used in medicine, ritual drinks and votive offerings. But nowadays due to its peculiar fixing properties, it is mainly used as a fixative in expensive perfumes. In United Arab Republic and other eastern countries, powdered ambergris is added to tea, coffee and other beverages for flavour. Owing to sporadic appearances, the exact price of ambergris is not known. However, the cost of ambergris may be about Rs. 25,000/kg (personal communication). Though the sperm whales are found distributed all over the seas, the ambergris is likely to appear only in oceanic regions where simultaneous abundance of sperm whales and their food namely squids and octopi is possible. The rare appearance of ambergris as noted earlier in the shallow coastal Devipatnam region in Tanjore District (Tamil Nadu) was due to a strong wind which could have washed the ambergris ashore.

Courtesy : Science Reporter

53

Multifarious Uses of Algae

It is estimated that aquatic algae synthesize 90 per cent of the world's organic carbon as they are known for the efficient utilization of solar energy than their counter part viz. higher plants.

Freshwater Algae in the Nutrition of Man and Live Stock

The potential value of freshwater algae as producers of concentrated protein, carbohydrate and fat is very great as there is no waste in the form of fibrous or woody portions, unlike in land plants in these algae. Many species of algae also serve as valuable sources of vitamins, hormones and antibiotics. The commonly occurring *chlorella* contains (by dry weight) 50% protein, 20% carbohydrate, 20% fat, and 10% ash and a number of vitamins, minerals and essential aminoacids and hence it may be processed to serve as a valuable ingredient in the human and live stock food.

Industrial Uses of Marine Algae

Marine algal resources especially seaweeds of India serve as valuable raw materials in the manufacture of commercially important products such as Sodium alginate, agar agar, iodine and various food products. While *Gelidium* and red algae are used in the production of agar agar, the brown seaweeds especially *Sargassum, Laminaria* and *Ascophyllum* are widely employed in the manufacture of alginate and as fertiliser and a feed supplement for livestock.

Algae as Food for Fish

Algae, being the first link in the food chain, determine directly or indirectly, the kind and the amount of food available for the growing fish. While the planktonic crustaceans and rotifers prefer the smaller algae, youngones of trout, perch and minnows feed on the zooplankton, the production of which depends ultimately on the planktonic algae, viz. phytoplankton. Milkfish of brackishwater ponds feed mostly on the filamentous blue-green algae and the yellow-green algae i.e. diatoms. Owing to the desirable nutritional characteristics of algae, they may be incorporated as the chief component in the production of protein-rich pelleted fish food.

Role of Algae in the Waste Water Treatment

Majority of the algae are known to purify the waste water systems. The different species of algae could indicate the nature of progress in the oxidation of the sewage stabilization ponds. For example, if the effluent of a pond contains

---

Courtesy : Aquaworld

many Chlorella, then the pond can work in its optimum level. On the other hand, if it contains a mixed flora, the pond can work at a higher level. And, if *Chlamydomonas* is present in the mixed flora, it proliferates when most of the organic matter is decomposed to make the nutrients available following the high pH. *Chlamydomonas* in turn excretes organic compounds which again make the minerals available for the growth of other algal species. Algal species could also be utilized in the removal of heavy metals especially in industrial effluents. For example *Closterium acerosum* and *Chlorella variegata* are known to remove chromium and iron respectively in effluents. Similarly *Navicula minima, Diatoma vulgar* and *Chlorogonium euchlorum* purify waste water systems in which the discharge of hydrogen sulphide, oil and distillery wastes is a regular affair. By selecting and adopting particular strains of algae, it is thus possible to bring out, biological changes in industrial wastes which otherwise may be toxic or unsuitable for release into streams, lakes or marine waters.

Uses of Algal Culture in Space Capsules

Mass culture of algae, one of the world's major organic resources is an important aspect as it serves as a chief food source for growing the larvae and adults of the commercially important finfish and shellfish in aquaculture practices. Mass culture of algae may be undertaken in closed systems of the space capsules. The nitrogenous wastes of the travellers could be diluted and circulated as algal nutrients and the respired air as a source of carbondioxide. In exchange, the travellers could utilize the oxygen released by the photosynthesis of algae and benefit favour the food value of cultured algae.

Algae as Indicators of Marine Pollution

In the polluted areas of marine environment, the sewage discharge stimulates the proliferation of algae such as *Enteromorpha, Ulva, Porphyra* and *Calothrix*. The red tide causing organisms such as *Trichodesmium erythraeum* (blue-green algae), *Noctiluca miliaris, Gymnodinimum breve* and *Gonyaulax* spp. (dinoflagellates) by virtue of their fast proliferation, may pollute the seas and deplete the oxygen content which in turn may lead to mass mortality of fishes during summer especially in the tropical seas.

Algae as Hydrobiological Indicators

The type of algae found in a particular stream may be a good indication of the range of temperature that the water has experienced. The optimum temperature for the diatoms has been found to be 18-30°C, for green algae it is 30-35°C, and for blue-green algae, 35-40°C. The diatom *Nitzschia linearis* which grows best at 22°C shows little or no growth at moderately higher temperatures. On the other hand, some species of blue-green algae are capable of growing at temperatures much higher than 40°C. For example, the optimum temperature for the blue-green algal species, *Oscillatoria filiformis* has been reported to be 45.2°C. Though diatoms have been recorded in moderately hot waters (40°C), green algae do not make their presence in hot springs.

Changes in the pH of the water due to industrial wastes or natural causes could also greatly modify the algal population. As mine discharge tends to lower the pH of freshwater drastically, the acid tolerant forms such as *Euglena mutabilis, Cryptomonas erosa* and *Ulothrix zonata* flourish in these environments. In cultures, the optimum pH for *Anacystis* (*Microcystis*) and *Cocchloris* has been found to be 10 and no growth of these forms has been reported in waters where the pH is less than 8.

Algae as Indicators of Toxin

Iron discharged from the steel mills is considered to be toxic to many algal species. However, the flageliates *Chromulina* spp and *Trachelomonas hispida* and diatoms such as *Pinnularta* sp. and *Eunotia* sp. could remain active and flourish in such environments owing to their tolerance towards higher concentrations of iron.

Further, algae generally require a small amount of iron for the production of their chlorophyll and most of the algae grow best when the iron content of their medium is between 0.2 and 2.0 mg/1. Distinct iron toxicity, however, is reported when its concentration exceeds 5.0 mg. Phenols at a concentration of 1.9mg/1 appear to be safe for diatoms in freshwater systems. Though copper is a well known toxin to many species of algae, species such as *Cymbella ventricosa, Calothrix braunil* and *Scenedesmus obliquus* have been reported to be resistant in moderate concentrations of this heavy metal. It is also worth-mentioning that *Spondylomorum pediastrum* and *Pandorina* sp. are capable of proliferating in paper mill wastes which are otherwise toxic to many species of algae. Hydrogen sulphide at concentration of about 4mg/1 has been found to be toxic to many species of diatoms which in turn affect the fisheries of an ecosystem. Hence, periodic monitoring of unwanted gases and heavy meals in aquatic systems is indispensable with a view to increase the fish production.

54

Prospects of Indian Seaweeds

Seaweeds include all macroscopic plants of the divisions algae viz. Chlorophyta, Phaeophyta, Rhodophyta and Cyanophyta occurring in the littoral and sublittoral habitats all along the Indian coast. Rich seaweed resources have however been recorded in the coasts of Bombay, Karwar, Ratnagiri, Goa, Varkala, Kovalam, Vizhinjam and Visakhapatnam. The seaweeds, contribute indirectly to the major chunk of the primary production of shore and continental shelf besides serving as good shelter for many species of fishes and invertebrates. The direct uses of seaweeds in the production of many commercially important materials are commendable.

Agar-agar

It is basically a carbohydrate and is a mixture of two polysaccharides - agarose and agaropectin. Certain species of red algae such as *Gelidiella acerosa, Gracilaria edulis, G. crassa G. verrucosa, G. corticata* and *G. folifera* are used in the extraction of agar-agar. The collected and cleaned weeds are bleached and dried in the sun. They are then ground to a pulp and dispersed in water. After allowing this to stand overnight, the supernatant clear liquid is decanted off and the residue is dried. The residue is then boiled with water, filtered hot and is allowed undisturbed to make the unwanted materials settle at the bottom. The clear supernatant solution is decanted off and is cooled when it jellifies. The jelly is cut into strips and dried. The agar-agar yield is estimated to be about 50 per cent of the dry weight of the sea weeds. The agar gel has got a peculiar property of setting it a fairly low temperature and the re-melting of the jelly once set requires a much higher temperature. This renders it highly suitable for inclusion in several food materials like ice cream, jams, jellies, salads, etc. A concentration of 2.2 per cent by weight of agar-agar added to jams and jellies in the last stages of their preparation gives them a fine consistency. Agar-agar can also be used in the preparation of sweets like Halwa and Payasam for giving them the desired consistencies.

Algin

Algin or alginic acid is the main polysaccharide occurring in the cell walls of brown algae. It consists of D-mannuronic acid and 2-glutaranic acid in various proportions. The brown algal species like *Sargassum, Turbinaria, Dictyota, Padina, Cystosetra, Hormophysa, Colpomenia, Spatoglossum,* and *Stoechospermism* are the good sources for the extraction of the commercially important algin. Algin and

---

Courtesy : Aquawori

its salts with sodium, ammonium, calcium, aluminium and copper have many industrial uses. The alginic acid can be prepared as follows.

Cleaned algin-yielding seaweeds are dried and boiled in water. The supernatant liquid is drained away and the residue is boiled with dilute hydrochloric acid. The clear portion is again decanted off and the residue is extracted repeatedly with dilute sodium hydroxide solution in the cold. All the alkaline extracts are combined and acidified and are treated with a dilute potassium permanganate solution for bleaching the product. The precipitated alginic acid is filtered and dried. Sodium alginate is used in the preparation of several food materials like cheese, custard, cake, jam, squash, etc., as a thickener and stabiliser. Dilute solutions of sodium alginate are known to be excellent glazes for frozen fishery products during storage. Calcium and aluminium alginates are useful in sizing and water proofing of fabrics. They also find use in the preparation of food, pharmaceuticals, rubber products, paper products, leather industry, cosmetics, shoe polishes, adhesives, textile products and photographic films.

As the copper aliginate is a toxic compound, it may be incorporated as an ingredient in the antifouling paints usable for fishing crafts in the prevention of damage causing boring and fouling organisms.

Seaweed as Human Food

High protein content of seaweeds especially of the rhodophytes, has all essential aminoacids, and rich sulphur and trace elements such as potassium, copper, magnesium, iron, manganese, etc., make seaweeds more suitable in the nutrition of human beings and livestock. It is interesting to note that the brown alga *Sargassum myriocystum* contains vitamin C at 70 mg/100g of fresh weeds which is higher than in the juice of lemon. Dried and cleaned seaweeds of *Gracilaria edulis* are used for porridges by the coastal populations of Tamil Nadu.

Seaweeds in Live Stock Feeds

Seaweed meal can be prepared by grinding the cleaned and dried seaweeds such as *Sargassum, Gracilaria, Ulva, Enteromorpha, Padina, Dictyota* and *Hypnea*. A mixture of seaweed meal and fish meal is a very good source of protein-rich live stock feed. *Garcilaria* meal is known to be a substitute for yellow maize and its inclusion at 10 per cent in the ration of white leghorns has yielded better growth and egg production. The residues obtained in the extraction of agar-agar can also be dried and pulverised for use as a supplementary feed for live stock owing to their rich trace elements.

Seaweeds as Manure

In India, seaweeds are nowadays used as a manure for the coconut plantations especially in the coastal areas of Tamil Nadu and Kerala. The water soluble potash, other minerals and trace elements present abundantly in seaweeds are readily absorbed by plants. Brown seaweeds owing to their algin content help in the conditioning soil facilitating aeration, moisture retention and adsorption of nutrient elements. Extracts of *Spatoglossum asperum, Ulva fasciata* and *Enteromorpha intestinalis* have been found to promote germination in seeds and growth of seedlings of gram, groundnut and maize. Liquid seaweed fertiliser is used as a leafy spray for better yields of various fruit trees. Composited manures using seaweeds and cowdung have showed better growth in plants like brinjal, tapioca, beans, greens, gourds, lime, papaya and drumstick.

Seaweeds as Fuel

Biogas plants can be installed using seaweeds as a raw material to produce methane. Addition to a specific strain of bacteria to seaweed digest is known to improve the yield of biogas.

Seaweeds in pharmaceuticals

Various species of red algae such as *Hypnea muserformis, Corallina officinalis, C. rubens* and *Alsidium helminthocorton* serve as vermifuges. *Gracilaria* and *Gelidium* are used in the treatment of various stomach ache and intestinal disorders. Carrageenan, an extract from red algae *Hypnea* and *Garcilaria* is useful in ulcer therapy. The alginates are found to prolong the rate of activity of certain drugs. Species of *Sargassum* are used for cooling and blood cleaning. Many species of seaweeds are rich in iodine. It is interesting to know that the green algae *Cladophora* contains 60 mg of iodine, the red *Gracilaria foliifera, Asparagopsis* sp. and *Polysiphonia* sp. contains 15 mg, 450 mg, and 40 mg, respectively and the brown *Sargassum* has 25 mg for every 100 g algae by dry weight. This paves way for utilising these seaweeds for large scale production of iodine in the production of medicine for treating goitre disease.

In recent years attempts have also been made to develop antimicrobial, antifungal and antiviral substances from seaweeds. Extracts of *Chondrus crispus* and *Gelidium cartilagineum* have been found to be active against influenza B and mumps virus. The extracts of *Enteromorpha* is known to be an effective antitubercular agent. *Padina terastromatica, Gelidiella acerosa* and *Acanthophora spicifera* exhibit 100% antifertility (anti-implantation) property in mice.

55

Seaweeds as Fertilisers

Seaweeds belonging to a number of species are found distributed all along the rocky areas of the east and west coasts of India. However, coasts of Bombay, Karwar, Ratnagiri, Goa, Varkala, Kovalam, Vizhinjam and Visakhapatnam are rich in sea weeds. The coral Islands of the Lakshadweep and Andaman also possess rich grounds for several varieties of seaweeds. The important genera occurring in our coasts are *Gracilaria, Hypnea, Sargassuum, Gelidium, Enteromorpha, Ulva, Halymenia, Sarconema, Turbinaria, Padina* and *Caulerpa.*

Many species of seaweeds have now been used as good sources of fertilisers especially in coastal localities owing to their mineral and trace element contents. Seaweed species like *Enteromorpha intestinalis, Padina australis* and *Gracilaria edulis* contain sodium, potassium, calcium, magnesium and chloride at 0.20 to 1.28, 0.63 to 2.01, 0.11 to 0.65, 0.16 to 0.50 and 1.25 to 3.70 (mg/100mg dry weight) respectively. In *Cladophora monumentalis, Ulva lactuca* and *Padina gymnospora* trace elements such as cobalt, copper, iron, manganese and phosphorus are present at 0.0125 to 0.1021, 0.5436 to 1.955, 0.3691 to 456.1, 6.152 to 24.75 and 28.625 to 277.6 mg/100g respectively.

The residue after the extraction of agar agar in red seaweeds is dried and pulverised and used as a manure for coconut plantations in the coastal areas of Tamil Nadu and Kerala. It is worthy of mention here that the water soluble potash and the aforesaid trace elements present abundantly in seaweeds are readily absorbed by plants. Brown seaweeds due to their algin content help in the conditioning of soil, facilitating aeration, moisture retention and absorption of nutrient elements. Extracts of *Spatoglossum asperum, Ulva fasciata* and *Enteromorpha intestinalis* are known to promote germination in seeds and growth of seedlings of agricultural crops such as gram, groundnut and maize. In order to increase the yields of various fruit trees, liquid seaweed fertiliser could be used as a leafy spray. Composite manures prepared out of seaweeds and cowdung have also showed better growth in brinjal, tapioca, beans, greens, gourds, lime, papaya and drumstick.

Courtesy : Shipping & Marine Industries Journal

56

Dunaliella—A Unique Halophilic Microalga

The importance of macroscopic seaweeds as food and source materials for industries is known from time immemorial. Similarly, the microalgae such as *Chlorella* and *Spirulina* have also been cultured in commercial scale for utilizing them in sewage treatment; as a source of fertiliser and as a protein-rich ingredient in animal and human foods. However, information on an extremely halophilic green alga *Dunaliella salina* (= *D. bardawil*) (Fig. 150) (Chlorophycea: Volvocales) is scanty. As this species is known to contain high levels of beta-carotene and glycerol, the large scale farming of this species is considered to be a profitable venture.

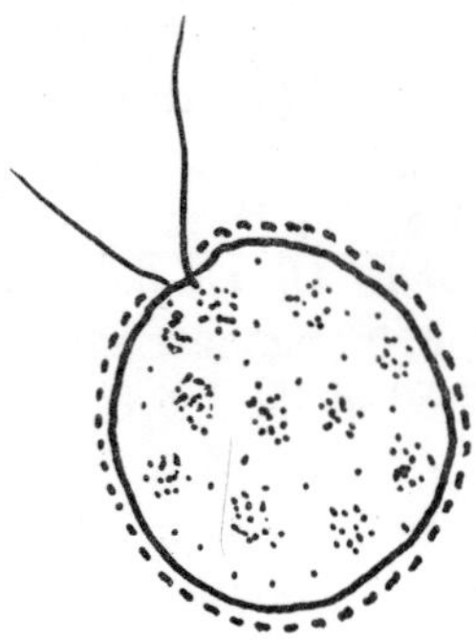

150

Biology of D. salina: *D. salina*, a unicellular organism has 2 flagella of equal length and a single cup shaped chloroplast like *Chlamydomonas*. The cell which is without a cell wall is motile even while dividing. *D. salina* forms resting spores (cysts) which are resistant to cold, rapid drying, nutrient depletion, etc. The cysts release single cells when they germinate. Although the shape of the cells may be oval, spherical, cylindrical, ellipsoidal, egg-pear or spindle shaped with radial, bilateral or dorsiventral symmetry or asymmetrical, the cells may change their shape with changing conditions and may become spherical under unfavourable conditions. Further the cell size may also vary with growth and light intensity. *D.salina* is also known to exhibit tolerance to DDT and uses various sugars in its osmoregulation. The pH tolerance range of this species is 6 to 11.

Distribution: *D. salina* occurs in salt marshes, salt pans and tide pools. The blooms of *D. salina* may colour the hypersaline water red.

Culture: As *D. salina* is extremely salt tolerant it can be grown in the brines up to 10 times saltier than normal sea water. Salt pan waters of 150-320 p.p.t. enriched with nitrogen, phosphorus and trace elements offer vast scope for the farming of *D. salina*. It is also known that the fast growth of this species is associated with 250 p.p.t. Under optimal environmental conditions, carbondioxide, sunlight and nutrients, *D. salina* is known to double its quantity in every two or three days. Further as predatory organisms such as protozoans and rotifers are absent in salt pan waters, *D.salina* cultures are extremely pure. The harvesting of this species is based on its hydrophobic properties i.e. the algae would adhere to surfaces which are also hydrophobic.

Vitamins and pigments: The cells of *D. salina* are known to contain a number of vitamins such as thiamine, pyridoxine, riboflavin, nicotinic acid, biotin and tocopherol in considerable quantities. Further this species contains many commercially important food pigments such as beta-carotene, xanthophylls (oxygenated carotenoids) and chlorophyllins. It is worthy of mention here that *D. salina* is the richest algal source of beta-carotene.

Extraction and uses of beta-carotene: The extraction procedure of beta-carotene from *D. salina* is a closely guarded secret in many west Australian companies who are the pioneers in the commercial cultivation and marketing of this species. In the extraction of beta-carotene, the filter harvested biomass of *D. salina* is subjected to solvent extraction. However, alternatively, the harvested algae could be freeze-dried to form algal powder which is rich in beta-carotene. This powder can be incorporated in livestock feeds to increase the fertility of pigs and cattle and also to increase the yellow colouring of eggs from battery reared chickens as well as giving a natural colour to the flesh of intensively reared trout, pigs, chickens and calves.

The extracted beta-carotene of *D. salina* is also used in the production of tablets for human consumption. It is known that the beta-carotene when stored in the fat tissues of human body it is converted into vitamin A as and when necessary. Unlike artificially manufactured vitamin tablets, the beta-carotene containing tablets do not give side effects in human beings. Natural beta-carotene of *D. salina* is also found to be anticarcinogenic and it may even be useful in treating cancer victims. Therefore the demand for the natural beta-carotene, the precursor of vitamin A of *D. salina* is increasing year after year. The present world market for this pigment is around 50 to 100 tonnes and is mostly provided by a petroleum derivative. One kilogram of synthetic beta-carotene costs about Rs. 15,000. The companies dealing with the beta-carotene sell their products either as solutions of 1.5 to 30% beta-carotene in vegetable oil; concentrated algal pastes or as spray dried algal powder.

Essential fatty acids: Unsaturated lipids present in *D. salina* can also be fractioned in order to produce more than five types of long chain poly and unsaturated fatty acids. Therefore, *D. salina* can also be profitably incorporated in aquaculture feeds.

Glycerol: *D. salina* constitutes about 80% glycerol in its dryweight. It is known that this species produces glycerol in response to higher salinities.

Other uses of *D. salina:* Extracts of *D. salina* can also be used as soil conditioners and fertilisers. Further, this species can be used to improve the efficiency of solar energy collecting systems.

57

Profitable Uses of Freshwater Weeds

Freshwater weeds are the highly productive macrophytes which require no tillage, seed or fertilisation. Owing to their excessive growth in the different waterbodies, there are two views as to whether to eradicate them or cultivate them. However, of late, these macrophytes have been found to be beneficial and hence the culture and propagation of these weeds are imperative.

The important freshwater weeds include emergent species like *Typha*, submerged species like *Potamogeton* and *Myriophyllum* and floating species like *Pistia, Eichhornia* and *Lemna,* of which the growth of *Eichhornia,* the water hyacinth is notable in every freshwater body and a production of even 200 tonnes per hectare is possible due to its vegetative reproduction.

The higher water content (75 to 95 per cent) of these weeds is a unique feature of aquatic weeds. Although the protein content of different weeds varies greatly (6 to 43 per cent), the amminoacid composition of many species is relatively constant, nutritionally well-balanced and similar to that of many forage crops.

The crude protein content of water hyacinth is 33 per cent which is comparable to that of soybean and cotton seed meal. Hence it will be lucrative to extract the leaf protein from these macrophytes for incorporating in the food of livestock.

The leaves of *Marsilea quadrifolia, M. crenata,* and *Pistia statiotes* and the stem of water lilies such as *Nymphaea lotus* and *N. stellata* are used as vegetables. The various parts of *Nelumbo nucifera,* the sacred lotus, are also used in a variety of cooked and fresh dishes. Similarly the rhizomes of *Potamogeton* and *Typha angustifolia* and the nut or kernel of water chestnut *Trapa* are also eaten in many parts of our country.

Hydrilla, Ceratophyllum, Najas and *Lemna* serve as important food for various herbivorous fishes such as grasscarp. Tilapia, gourami, Puntius, commoncarp and milkfish. It is very interesting to note that the grasscarp can consume aquatic weeds at the rate of 170 per cent of its body weight every day and can grow to a maximum of 5 kg in an annum. Further this living manuring machine is known to send out 50 per cent of the consumed food in undigested condition which inturn fertilises the fish ponds.

Freshwater weeds can also be a good food source for other herbivorous animals such as turtles and rodents particularly rabbits.

The higher water content and mineral content seem to reduce the palatability of livestock over these macrophytes. However, species of *Pistia, Lemna, Typha, Nymphaea, Hydrilla Ceratophyllum* and *Eichhornia* can be directly used as pig and

---

Courtesy : The Hindu

duck feed. A combination of partially sundried or dehydrated *Eichhornia* (10 kg), 1.4 kg of paddy straw and 0.7 kg of groundnut cake will be an ideal buffalo fodder.

Freshwater weeds may serve as organic fertilisers, green manure or compost. *Pistia, Hydrilla, Salvinia* and *Eichhornia* serve as good manure due to their high potash content. Similarly the application of the fern *Azolla* along with the nitrogen fixing symbiotic bluegreen alga viz. *Anabaena azollae* has been found to enhance the rice yield. Further application of compost prepared by decaying weeds along with soil, ash and animal and human waste to fish ponds will not only enhance the growth of fish food organisms but will also curtail the expenditure towards fertilisation of fish ponds.

Though animal manure or human wastes mixed with vegetable matter serve as biogas digesters in many countries, it is now understood that *Eichhornia* can also be directly digested to produce methane due to its carbon and nitrogen ratio (30:1). It is reported that 1 kg of *Eichhornia* can yield 370 l of biogas. The resulting biogas slurry, which is without carbon but with nitrogen can be used as organic fertilisers particularly for fish ponds.

Species like *E. crassipes* can be combined with banana stem in the preparation of paper pulp. (Fig. 151)

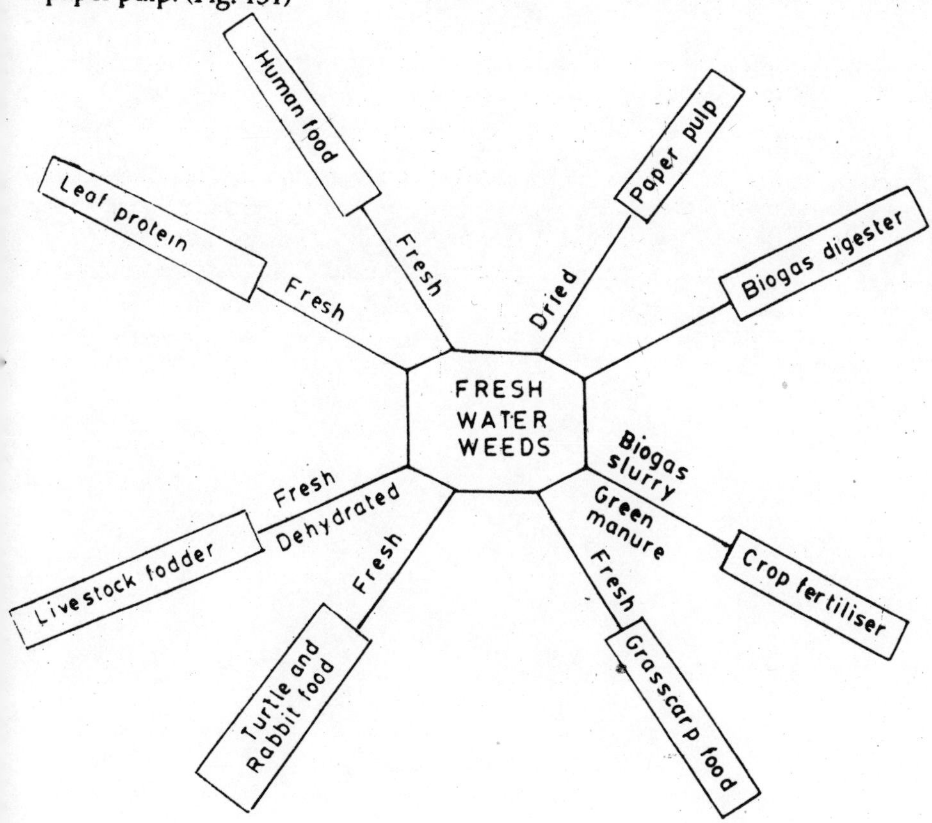

151

Index

Acanthophora Spicifera 160
Acropora 108
Actinomycete 120
Active fishing gear 128
Agar agar 155, 158, 159, 161
Agar gel 158
Agaropectin 158
Agarose 158
Agglomerated lorica 94
Air bladder 148
Air breathing fish culture 1
Air lift 63
Air stone 64
Algal bloom 120
Algal paste 163
Algal powder 163
Algin 158, 159
Alginic acid 159
Alkali digestion 146
Alpheidae 114
Alpheus euphosyne 114
A. malabaricus 114
Alsidium helminthocorton 160
Alternating current 128, 130, 132
Ambergris 154
Ambreine 154
Anabaena azollae 53, 165
Anabas testudineus 8
Anaesthetics 76
Anchor worm 71
Anchovy 89, 146, 152
Annular chamber 126
Antarctic krill 91, 92
Anthopleura xanthogrammica 84
Anthopleurin-A 84
Antibiotics 84, 155
Antiboring paint 103
Anticlotting agent 84
Anticoagulant 84
Antifertility 160
Antifouling paint 103, 105
Antifungal substance 160
Antipatharia 108
Antiimplantation 160
Antimicrobial substance 160
Antitubercular agent 160
Antiviral substance 160
Anticarcinogenic substance 163
Aplocneilus blochi 2
Aquaculture 1, 17, 53, 132
Aquatic fern 53
Arabinosyl cytosine 84
Argulus 71
Arrow worm 91, 92
Artemia 74
Arthritis 122
Artificial drier 148
Artificial light 130
Artificial park 139
Artificial seawater 82
Ascophyllum 155
Asparagopsis 160
Asterionella japonica
Atoll 106
Attractant 139
Azeotropic mixture 147
Azolla pinnata 53
Bacilla rubra 51
Baleen whale 91
Barbus 69
Barrier reef 106
Bathyoplectidae 123

Beche-de-mer 116, 117, 146
Bekti 119
Beta-carotene 162, 163
Betta tail 58
Bheries 3, 30
Biological filter 49, 62-64
Black coral 108
Blazka apparatus 125, 127
Blood worm 46, 65
Blowfish 121
Bottom sword tail 58
Bottom trawling 130
Bouchot culture 33
Brachionus angularis 97
B. calciflorus 97
B. Plicatilis 74, 75, 99
B. quadridentatus 99
B. rubens 97
B. urceolaris 97
Breeding hapa 57
Brine shrimp 65, 83
Brown mussel 39, 42
Bubble float 100
Cabinet 134
Cage 36
Calanus 91
Calicoblastic layer 106, 108
Caligus 71
Calothrix 156
C. ventricosa 157
Calpomenia 158
Carangid 139
Cardiac stimulant 84
Catalytic agent 147
Catamaran 139
Catfish 16, 57, 119, 139, 148
Catla 1, 4, 5, 8, 11, 16, 21, 57, 120
Caulerpa 43, 161
C. racemosa 43
Cavolinia 101
Ceratophyllum 167
Chaetognatha 96
Channa marulius 8
C. striatus 8
Chanos Chanos 24
Chinese carp 21
Chinese dipnet 130
Chitin 84
Chitosan 84
Chlamydomonas 156, 162
Chironomus 46
Chlorella variegata 156
Chlorogonium euchlorum 156
Chlorophyllins 163
Chlorophyta 162
Chloroplast 162
Chondrus crispus 160
Chromulina 157
Cichlid 71
Ciliata 94
Circular net 130
Cirrhina mrigala 4
Cladocera 75
Cladophora monumentalis 161
Clarias macrocephalus 8
Clione 100, 102
Closterium acerosum 156
Coastal upwelling 89
Cocchloris 157
Coccolithophore 88, 90, 96
Cocconeis ceticola 86
Cofer tail 58
Cold living light 91
Composite fish culture 1, 4, 6
Common carp 1, 4, 6, 9, 11, 16, 120
Congiopodus spinifer 123
Contact plate freezer 148
Corallum rubrum 108
Corallite 108
Coral reef 106, 11
Coral sea 108
Costia necatrix 69
Costiasis 69
Cotton seed meal 164
Cotton wool disease 69
Craft 128
Crassostrea cucullata 31
C. discoidea 31
C. gryphoides 31
C. madrasensis 31, 33
Creseis 101
Crop fertilizer 165
Cryptomonas erosa 157
Cryopreservation 80, 81
Cryoprotectant 80, 81
Crystalline guanin 149
Ctenopharyngodon idella 4, 8
Cyanophyta 158
Cyclops 22, 65
Cymbella ventricosa 157

Cyprid 22
Cyprinodontidae 58
Cyprinus carpio 4, 8
Cystosetra 158
Dactylogyrus 71
Danio 69
Daphnia 22, 65, 74, 75
DDT 162
Deacetylisation 145
Delta tail 58
Demersal fish 139
Demineralisation 145, 150
Deproteinisation 145, 150
Deskinning 146
Detritus 143
Diatom 17, 86, 96, 155, 157
Diatomaceous earth 86, 91
Diatoma vulgar 156
Diatomin 86, 91
Diatomite 86
Dictyota 158, 159
Digitalis 84
Diluent 80
Dimmerstat 125
Dinoflagellate 85, 88, 91, 92, 130, 156
Dipnet 130
Dipnet fiahing 130
Direct current 128, 129
Double sword tail, 58
Dragged fishing seine 128
Drag seine 133
Dropsy 61, 69
Dry pellet 65
Dry pelleted feed 68
Duck-cum-fish culture 2
Dunaliella bardawil 162
D. salina 75, 162, 163
Echinophora 108
Echobuck 135
Echogram 135, 136, 138
Echomark 136, 138
Echosounder 134, 136, 138
Ectoparasite 73
Eichhornia crassipes 10, 16, 165
Elatol 84
Electric fish 130
Electric fish fence 132, 133
Electric harpoon 132
Electric organ 130
Electric whaling 132
Electrified fishing gear 132, 133
Electrified hook 132
Electrified seine 130
Electrocauterisation 29
Electronarcosis 133
Ensilage 149
Entagling net 139
Enteromorpha compressa 53
E. intestinalis 43, 53, 156, 159-161
Epicoprostanol 154
Euchaeta norvegica 92
Euglena mutabilis 157
Euphausia superba 91
Eunotia 157
Evil globe fish 121
Exotic carps 4, 17
Extender 80, 81
Eye stalk ablation 29
False bottom 62, 63
False coral 108
Fan tail 58
Feed pellet 68
Filina longiseta 97
Fillet 148
Fin rot 69
Fish activity counter 126
Fish aggregating device 139
Fish glue 147
Fish louse 71
Fish offal 149
Fish protein concentrate 92, 147
Fish pulp 147
Fish sauce 148, 152, 153
Fishery byproduct 145
Flag tail 58
Food chain 88, 96
Food conversion ratio 1
Fragilaria antarctica 86
F. oceanica 86, 90
Fry respirometer 126
Fungia 108
Gambusia affinis 2
Gas bubble disease 71
Gasoline 149
Gazani farm 3
Gelidiella acerosa 158, 160
G. verrucosa 158
Gelidium cartilagineum 160
Geosmin 120
Giant freshwater prawn 17

Giant gouramy 71
Glue liquor 147
Glycemia 128
Gonad inhibiting hormone 29
Gonad stimulating hormone 29
Gonopodium 58
Gonyaulax acatella 92
G. catenella 92
G. monilata 92
G. polyedra 92
G. tamarensis 92
Gracilaria corticata 53, 158
G.crassa 158
G.edulis 158, 159, 161
G. folifera 53, 158
Grass carp 1, 4, 5, 8, 9, 11, 21, 81, 120
Great barrier reef 111
Green mussel 39, 42
Gregor Mendel's law 58
Grey mullet 81
Gymnodinium preve 92, 156
Halobates germanus 112
H.micans 112
H.sericeus 112
H. sobrinus 112
H. splendens 112
Halophilic microalgae 162
Halymenia 161
Handline 139
Hand pelletiser 68
Handpicking 143
Harpoon gun 132
Hatching hapa 57
Heliopora 108
Hemidiscus hardmannianus 86, 90
Hemiptera 112
Heperin 84
Hexarthra fennica 99
Holoplankton 100
Holothuria argus 116
H. echinitis 116
H. edulis 116
H. lecanova 116
H. mauritiana 116
H. nobilis 116
H. scabra 116, 117, 146
Holothurin 117
Hormophysa 158
Human chorionic gonadotropin (HCG) 55, 57
Hydrilla 4, 164, 165
Hydrophobic 163
Hydroponics 51
Hypophthalmichthys molitrix 4, 8
Hypophysation 55
Ichthyophthirius 69
Index of spoilage 120
Indian major carps 17, 21, 57, 81
Indian salmon 119
Induced breeding 41, 132
Influenza B 160
Instant sea salt 83
Integrated fish farming 1, 7
Iron toxicity 157
Jelly fish 88, 92
Janthina 100
Keratella cochlearis 97
Khozan land 3
Kieselghur 86
King cobra tail 58
Kissing gouramy 71
Krill 86, 91, 92
Labeo rohita 4, 8
Lace tail 58
Laminaria 155
Larvicidal fish 2
Larvivorous fish culture 2
Leaf protein 165
Lebistes reticulatus 58
Lecane bulla 99
L. lunaris 99
Leiognathids 146, 152
Lemna 4, 164
Lepadella ovalis 99
Lesser sardines 146
Leucaena leucocephala 53
Lift net
Light fishing 130
Limacina 92, 100, 101
Limnoria 103
Liquefaction 149
Liquid seaweed fertilizer 160
Longline 128, 132, 139
Mackerel 139, 147, 149, 152
Macrobrachium rosenbergii 8, 17-20
M.malcomsonii 17, 19
Macrophyte 164
Maghur 1, 57
Mahua oilcake 27
Major element 82

Mangrove 3, 143
Mangrove leaves 53
Mariculture 36, 83
Marine algae 155
Marine fouler 103
Marine park 108
Marine pollution 156
Marsilia crenata 164
M.quadrifolia 164
Mass mortality 92
Mechanical feeder 65
Mediterranean skate 130
Medusae 112
Megalopa 29
Megalops cyprinoides 21
Mesogastropoda 100
Metapenaeus dobsonii 143
Metazoan parasite 69
Microbial contamination 152
Microcystis aeruginosa 97
Microflagellate 88
Microphone 134
Microzooplankton 89
Migratory fish 139
Milk fish 20, 24-26, 81, 155
Millions fish 58
Mimosine 53
Minnow 155
Molly 69
Monoculture 17
Monsoon river prawn 17
Montipora 108
Mooring boat 130
Mooring line 139
Moss animal 103
Mrigal 1, 4, 5, 11, 16, 21, 57, 120
Mud crab 28-30
Murrel 1, 11, 14, 16, 119
Mullet 20, 119, 149
Mussel 39, 40
Mutation 60
Myriophyllum 164
Najas 4, 164
Nanophytoplankter 33
Nanoplankton 88
Navicula minima 155
Nelumbo nucifera 164
Neoplasius 69
Neuralgia 122
Neurotoxin 85, 92
Niacin 118
Nicotinic acid 163
Nitrobacter 49, 62
Nitrosomonas 49, 62
Nitzschia linearis 156
Noctiluca miliaris 156
Nymphaea lotus 164
N.stellata 164
Oceanic insect 112
Oceanic skater 112
Offbottom culture 31
Oil sardine 90
Onbottom culture 31
Ooze 100
Opisthobranch 100
Organ pipe coral 111
Organs of Cuvier 117
Oreochromis mossambicus 21
Ornamental fish culture 3
Oryzias melastigma 2
Oscillatoria 120
O. filiformis 156
Oxidation pond 1, 10, 15
Padina 158, 159
P. australis 161
P. gymnospora 161
P. tetrastromatica 160
Pandorina 157
Paper pulp 165
Paper speed 136
Parthenogenesis 75
Passive fishing gear 128
Pearl essence 148
Pecten plica 36
P. splendidulus 36
P. tranquebaricus 36
Pelagic fish 139, 149
Pelleted fish food 155
Pen 33
Penaeus indicus 24, 143
P. monodon 24, 77, 143
P. semisulcatus 143
Pen speed 135, 136
Perch 155
Perna indica 39, 40
P. viridis 39-41
Phaeophyceae 158
Phagocyte 84
Phasing 135, 136
Phoenix petrel 112

Phosphate island 111
Phosphatisation 111
Photocell rack 126
Photocell system 126
Photochromic eye 121
Photosynthetic algae 111
Phytoplankton 4, 5, 10, 29, 39, 42, 86, 88, 120, 155
Pig-cum-fishculture 2
Pin tail 58
Pinnularia 157
Pistia statiotes 164
Pistol hand 114
Pistol shrimp 114
Pituitary extract 55
Planktonic ciliata 94
Planktonic mollusc 100
Platy 61
Poecilia reticulata 58
Pokkali field 3,30
Pole farming 33
Polyarthra vulgaris 99
Polynemid 148
Polyculture 17, 24
Polymorphism 36
Polysiphonia 160
Porites 108, 111
Porphyra 156
Potamogeton 164
Prawn picking 143
Primary production 88, 90
Pteropod 92, 100
Puffer 121
Pulsating current 128, 130, 132
Pulse length 135, 136
Pump fishing 132
Pyridoxine 163
Quinaldine 76
Quarantine tank 73
Rack 32
Radiolaria 94, 100
Raft 32, 139
Ray 119, 146
RBC 128
Red tide 92, 156
Reef fish 111
Rhincodon typus 90
Rhizosolenia 92
Rhodophyta 158
Ribbon fish 149, 152
Riboflavin 163
Robson round tail 58
Roe 118, 152, 153
Rohu 1, 4, 5, 11, 16, 21, 57, 81, 120
Rope culture 40
Rotatoria neptunia 97
R. rotatoria 97
Rotifer 22, 65
Round tail 58
Sacred lotus 164
Sagitta elegans 91
S. setosa 91
Saline solution 146
Salmonella 13
Salp 96
Saltpan water 162
Salvinia 165
Saprolegnia 69
Sarconema 161
Sardine fishing 130
Sardinella longiceps 146
Sargassum 43, 155, 158-161
S. ilicifolium 53
S. myriocystum 159
Scale carp 11
Scallop 36-38
Scenedesmus obliquus 157
Schroederella delicatula 89
Sciaenid 147, 148, 152
Scissors tail 58
Sclerosepta 108
Scylla serrata 28, 30
Sea anemone 112
Sea bass 148
Sea butterfly 100, 102
Sea cucumber 116, 117, 146
Sea fan 108
Sea feather 108
Sea flower 108
Sea slug 111
Sea squirt 103
Sea urchin 111
Sea weed digester 160
Sea whip 108
Secondary production 88, 89
Seer fish 119
Seine 128, 139
Selective breeding 80
Semiautomatic pelletiser 54
Sewage fed fish culture, 1, 15

Sexing 58
Shark 119, 139, 146
Shark fin 145
Shark fin soup 145
Shelf life 142, 148
Shellfish 83
Shigella flexneri 13
Ship worm 103
Siamese fighter 61, 69
Silicoflagellate 88, 96
Silvercarp 1, 4, 5, 8, 9, 11, 16, 21, 57, 81, 120
Singhi 1, 57
Skate 119, 146
Skeletal septa 108
Skeletonema costatum 89
Slow digestion 152
Sludge 15
Snapping shrimp 114
Sodium alginate 155, 159
Soil conditioner 163
Solar drier 141, 142
Solar drying 142
Solar radiation 142
Solar tent drier 141
Soles 119
Soluble matrix 105
Sound beam 135, 136, 138
Sound pulse 135
Sound strength 138
Sound wave 135
South American eel 130
Space capsule 155
Spade tail 58
Spat 32, 41
Spawner 132
Spat collector 32
Spatoglossum 158
S. asperum 160, 161
Spear tail 58
Sperm whale 154
Sphaeroma 103
Spirodella 4
Spirulina 75, 162
S. platensis 53
Spondylomorum pediastrum 157
Squid 156
Squid jig fishing 130
Standing boat 130
Stationery electric field 132
Stoechospermism 158
Stony coral 108
Subabul 53
Sun drying 142
Supplementary feeding 20
Suppression 138
Surrounding net 139
Swellfish 121
Swim bladder 148
Sword tail 61
Symbiosis 111
Symbiotic blue green algae 53, 165
Tagging 132, 133
Tail rot 69
Tarpon 21-23
Tetraodon 121
Tetradontidae 121
Tetradotoxin 121
Tetraselmis 75
Thelonota ananas 116
Thorny coral 108
Thryssa 152
Tilapia mossambica 8, 11
Tintinnid 94
Tocopherol 163
Topsword tail 58
Transducer 134
Trawl 128, 130
Trawling net 138
Trawl mouth 132
Trepang 116
Tricaine methane sulphonate (MS 222) 76
Trichodesmium erythraeum 156
Tri-n-butyltin fluoride (TBTO) 105
Tri-n-butyltin oxide (TBTF) 105
Trochelminthes 97
Trymethylamine 120
Trochocera similis 99
Trochus 111
Trophic level 89, 114
Trout 155
True coral 108
Tube worm 103
Tubipora 111
Tuna fishing 132
Turbinaria 43, 158, 161
Tuxedo tail 58
Typha angustifolia 164
Ulcer therapy 160
Ulva 43, 156, 159, 161
U. fasciata 160, 161

U. lactuca 161
Upwelling 88-90
Veiltail 58
Velvet disease 71
Vermifuge 160
Verongia gigantea 84
Volvocales 162
Warp 130
Water hyacinth 10, 164
Water recirculation system 49, 50
Wet feed 65
Wet reduction process 146
Whale shark 90
White line 138
White spot disease 69
White water 90
Wood borer 103
Xanthophyll 163
X organ sinus gland 29
Zero line 134, 138
Zeromark 134, 136
Zooplankton 4, 5, 10, 88, 91, 108, 120, 155
Zooxanthellae 111